V-2

A Combat History of the

First Ballistic Missile

T. D. Dungan

WESTHOLME
Yardley

First Westholme paperback 2019

Copyright © 2005 Tracy Dwayne Dungan

Westholme Publishing, LLC
904 Edgewood Road
Yardley, Pennsylvania 19067
Visit our Web site at www.westholmepublishing.com

ISBN 978-1-59416-327-2
Also available as an eBook

Printed in the United States of America on acid-free paper.

Contents

1. Beginnings   1

2. Pennemünde Research   27

3. Discovered   57

4. Preparations for Vengeance   83

5. Operation Penguin   115

6. Twilight of the Gods   136

7. City of Sudden Death   160

8. Final Retaliation   172

9. Legacy   205

Appendix   220

Notes   230

Sources and Bibliography   238

Index   243

Acknowledgements   250

1

Beginnings

Walter Dornberger glanced around the table. Seated with him were the critical players, all men that he had hand chosen. It was a night of celebration, the culmination of more than a decade of intense experimentation and struggle. That evening at Dornberger's home, the spirit of camaraderie was strong. Each man in attendance felt a sense of personal accomplishment. They had achieved something together, as a team, something heretofore unimaginable to even the most celebrated scientific minds of the day. Dornberger raised his glass to make a toast.

"For the first time we have invaded space with our rocket. Mark this well, we have used space as a bridge between two points on the earth; we have proven rocket propulsion practicable for space travel. This third day of October, 1942, is the first of a new era of transportation: that of space travel."[1]

One might imagine the part of the speech that evening that didn't make it into the history books—the part where Dornberger most probably spoke of the triumph for the Fatherland and Nazi Germany; the part about the creation of the ultimate weapon to serve the Führer and to bring victory to the German people by striking the enemy with swiftness and invulnerability. Walter Dornberger was a devoted Nazi officer. He believed in Hitler; he believed in the war. He also was a man with a vision. The drive, energy, and resourcefulness of General Dornberger and the ingenuity and passion of young rocket scientist Wernher von Braun combined to create one of the greatest technological achievements of the twentieth century. Along with thousands of other engineers and scientists, they produced one of the most infamous weapons of the Second World War: the V-2 rocket. But at what cost? The fact that this achievement was the product of the ruthless regime of Nazi Germany, brought to the battlefield by any means necessary, and the fact that many thousands of people died as a result of its production and deployment taint the image of what otherwise would be considered a monumental scientific triumph. The German scientists brought to the United States immediately following the end of the war in Europe have been praised as heroes for their contributions to America's postwar space exploration, especially the moon landings. However, many now believe the German rocket team was guilty of contributing greatly to the suffering of the slave laborers who were forced to assemble the V-2s while enduring inhuman treatment at the hands of the ruthless SS. Political aspects aside, the fact remains that these men shaped the ground rules for warfare in the later years of the twentieth century.

The monumental leaps in technical abilities during the last century have altered the landscape of the modern battlefield. The ability to conquer an enemy with overwhelming firepower along with the technological means to achieve it, propelled many sweeping endeavors in science and technology. Not only did military ambitiousness set the stage for new technologies, other advances in household conveniences, modern transportation, and even recreational activities were the by-products of building a more advanced military. The dream of supersonic flight and manned space exploration were infiltrated repeatedly for martial purposes.

The V-2 or A-4 (Aggregat 4) was the first long-range ballistic missile to be actively used in combat. This huge German missile hurtled a 1,700-pound warhead 50 miles high and hundreds of miles downrange to its target. There was no stopping it, no countermeasure to shoot it down. It traveled at supersonic speeds and slammed into the earth without the slightest warning. The V-2 was Germany's most monumental manufacturing endeavor of the war, but as with all Nazi secret weapons, its introduction was far too late to influence the final outcome of the war. It is arguable whether the missile could have made a difference if introduced earlier in the war. Even if tens of thousands could have been produced and launched against Allied cities and ports in the months prior to D-Day, the Soviet juggernaut was still marching forward in the east; Germany could not have deterred the Red Army with V-weapons.

Two years after the joyous celebration at Dornberger's domicile, Germany would find itself in a desperate position, battling for its very survival. By the fall of 1944, the brainchild of the German rocket scientists was raining destruction down on London and Antwerp. The Vergeltungswaffe Zwei (Vengeance Weapon 2) seemed to be Germany's last desperate hope. However, planning and production of the rocket weapon was extensive and wide ranging. Although delayed by Hitler's shortsightedness, it was not an operation thrown together at the last hour. It was the culmination of years of strategic thinking. Even though the weapon did not tip the balance in Hitler's favor, the mobile deployment of the V-2 hampered Allied war efforts until the very end. The German rocket has been criticized as too expensive, too complicated, and too inaccurate. While those attributes ring true, the detractors never take into account the unprecedented invulnerability of the missile and the tremendous influence it had on Allied war planning. How was it possible for the Germans to launch over 3,000 V-2s at Allied targets when their economy was seemingly in shambles? Why was V-2 production going strong, even until the last days of the war, despite the fact Germany was bombed into oblivion? Was it conceivable that even when Allied fighters and bombers owned the skies over Western Europe, not one mobile V-2 attack was ever prevented by Allied forces? And how would this fact influence strategic planning in the future?

When the United States dropped the first atomic bombs on Japan in 1945, the face of conventional aerial warfare changed forever. No longer would it be necessary to send armadas of bombers to destroy a city. One nuclear device, and the means to deliver it, would be the only prerequisite. Today, the single greatest strategic threat to any nation is an attack by one or more ballistic missiles armed with nuclear or other weapons of mass destruction. Many nations in the world today possess the ability to manufacture nuclear weapons. It is only those with the ability to deliver such weapons to their enemies that are considered a substantial threat to other nations. As the grandfather of all modern ballistic missiles, the German V-2 was armed with a conventional warhead and was only capable of causing significant damage if fired in large numbers against its target. Today's ballistic missile warheads can be single or multiple and can deliver nuclear, chemical, or biological payloads as well as conventional high explosives. As early as 1943, Adolf Hitler foresaw the present-day dilemma of the proliferation of such weapons and the coming ballistic missile threat when he predicted, "The world will be too small now to contain a war."

The Space Age began long before the Soviet Union's *Sputnik I* orbited the earth in 1957. This event was just one in the continuing series of rocketry endeavors going back more than 30 years. Much of the world's early experimentation in the field of rocketry was pioneered by amateur enthusiasts in Germany following the First World War. After the armistice of WWI, the victorious Allies were determined that Germany should not rise again to rekindle the conflict so freshly ended. With an aim to incapacitate their enemy's ability for future aggression, the Allies dissected the territory of the German empire and split it up between the surrounding countries. After the war, unemployment soared, inflation grew at a staggering rate, and insurrection in Germany began to emerge. The conditions in the Weimar Republic subsequently conditioned what followed—a second world war.

In 1923 unknown author Hermann Oberth published a small paperback book titled *Die Rakete zu den Planetenräumen (The Rocket into Planetary Space)*. He attempted to prove, by using scientific and mathematical evidence, that launching rockets into space was practical. Oberth was born in Transylvania and became a medical student at Munich. He had become greatly interested in the development of space vehicles. The book stirred the minds of its readers with the notion that with certain advancements in technology, manned spaceflight was possible in the near future. The book discussed real physics and aerodynamics of a rocket using liquid propellant (that could be throttled). It proposed an instrument-carrying rocket and a vehicle that might carry humankind into outer space.[2]

The scientific community took little notice of Oberth's book, and those that did offer commentary often ridiculed the book as being insufficient in scientific research. The book was somewhat nontechnical in many areas, and it was for this very reason that it quickly became popular with the average enthusiast in the Weimar Republic. It became popular enough that there was a second printing in 1925.

In 1926 German interest in space travel waned. The educational and technical requirements to pursue these studies limited the number of persons who could actually participate in real research. While many German citizens entertained magnificent visions of gleaming spaceships, very few could understand the mathematics contained in such scientific pamphlets. It was at this time that sensationalist Max Valier came upon the scene. Valier had become caught up in the fever spurred by the writings of Oberth and others. Valier wrote numerous scientific articles for magazines and journals of the day. In 1924 he wrote *Der Vorstross in Weltraum* (*The Drive to Outer Space*), a book capitalizing on Oberth's success of a few months earlier. Even less scientific than Oberth's work, Valier's writing contained many errors. However, the excitement in his writing and the nontechnical language he used made it even more appealing to the average space enthusiast. Growing quickly in popularity, he soon would propose rocket-powered cars, railcars, and gliders. Along with collaborators such as Fritz von Opel (of German Opel cars), Valier built rocket-powered experimental vehicles of many varieties. His publicity stunts drew hundreds of spectators to see the huge plumes of white smoke and hear the deafening roar of solid-propellant rocket motors. Around 1930 Valier started experimenting with liquid-fuel rocket engines. He was killed on May 17, 1930, during an experimental engine test in Berlin. The rocket engine combustion chamber exploded, and shrapnel pierced his heart.

In 1926 the book *Die fahrt ins Weltall* (*Journey into Space*) by Willy Ley was published. Ley's writings were based in reality and offered good explanations into rocketry theory. However, Ley was unable to garner public interest while the charismatic Valier was traveling around the country amazing the masses with his rocket-powered vehicle exhibitions. On July 5, 1927, while sitting in the back of a Breslau restaurant, a group of engineers, theorists, and science students formed an association to conduct real research into rocket design and applications. Tired of the publicity stunts, they formed the Verein für Raumschiffahrt, or VfR (Society for Space Travel).

The VfR grew in popularity and rapidly increased its membership to almost 800. The stated purpose of the society was to conduct rocketry experiments, but early on the group actually did very little in this capacity. During the society's monthly meetings, much discussion, arguing, and brainstorming preceded a short meeting, which was followed by the distribution of the club's newsletter called *Die Rakete* (*The Rocket*). In addition to Ley, the VfR membership included such notables as Hermann Oberth, Johannes Winkler, Dr. Walter Hohmann, and (not so notable at the time) a teenager named Wernher von Braun.[3]

In 1929 Rudolf Nebel joined the VfR society at the urging of Willy Ley. Nebel had been involved with Oberth on an earlier failed project, and since that time, he and Oberth had been on bad terms. There were some tense moments at the next few club meetings as the two men offered differing opinions on how to go forward with the development of a liquid-fuel rocket. Nebel was not a scientist, but he brought a practical engineering viewpoint to the rocketry discussions of the VfR. Believing it best to begin with the basics, he proposed a small liquid-rocket design. The proposal was accepted by the membership with the exception of Oberth, who felt the design too faint. Nevertheless, work moved forward on the project called Mirak. The society was at first very excited to seek publicity for their upcoming Mirak engine tests. However, that summer, after the tragic death of Max Valier, public opinion concerning rocketry changed somewhat, and the group decided to conduct their trials in private. Nebel and Klaus Riedel moved to a farm in Saxony, away from view, to conduct the rocket tests. Reports detailing the test results were published and distributed to the VfR membership. Society members waited anxiously for word of the next successful firing or unexpected explosion.

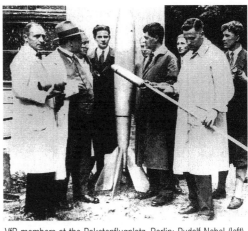

VfR members at the Raketenflugplatz, Berlin; Rudolf Nebel (left), Hermann Oberth (next to large rocket in long coat), Klaus Riedel (holding small rocket), and a young Wernher von Braun (behind Riedel). (*NASA*)

By this time, it was plainly obvious that the VfR needed a better location to conduct their experiments. The not-so-rural open field they had been using at Bernstadt suddenly seemed inadequate. It was in late 1930 when the society happened upon a deal they could not pass up. In the northern Berlin suburb of Reinickendorf, Nebel located an abandoned ammunition storage complex, four square kilometers in size, complete with roads and buildings. The society was able to rent the complex from the municipality of Berlin for the modest sum of ten reichsmarks annually. The VfR's Raketenflugplatz (rocket airport) was opened on September 27, 1930. By March of the following year, the site was ready for operation. Many improvements had been made to the facility along with the construction of a basic test stand for static firings.

In May of 1931 Klaus Riedel designed a new rocket, the Mirak II or Repulsor series, using the thrust chamber developed for the Mirak, fed by two long tanks containing liquid oxygen and gasoline, which would form guiding sticks for the forward-mounted engine. Test results were so encouraging that some in the group were talking about the possibility of actually launching a version of their new one-stick Repulsor rockets. On May 10, 1931, Riedel was alone at the Raketenflugplatz running tests on the flying variant of the design, when suddenly, to his surprise, the rocket lifted slowly and rose to about 18 meters. Then the motor shut off, and it fell to the ground, damaging it slightly. The Repulsor was repaired and on May 14, 1931, made its first official flight.[4]

Across the ocean in the United States, the enthusiasm for rocketry was not as prevalent as it was in Europe. Only one man, Dr. Robert Goddard, had conducted extensive experiments with liquid fuels applied to rocket propulsion. As early as 1923, Goddard was having success with liquid-fuel rocket engines. In 1926 he launched one of his creations to an altitude of 41 feet in a flight that lasted a little over two seconds. With the help of Charles Lindbergh, Goddard was able to secure a sizable research grant, and he moved to the vast open spaces near Roswell, New Mexico, to continue his research.

Throughout the early 1930s, he built larger and more complex rockets while developing many of the same ideas that would be mirrored in Germany.

In 1930 Goddard was approached by a newly formed group of enthusiasts based in New York City asking him to publicize his work. After he refused, two founding members of the American Interplanetary Society traveled to Germany in the spring of 1931 to make contact with the VfR. The Americans were warmly greeted by Willy Ley who gave them an extensive guided tour of the Raketenflugplatz in Berlin.

The first year of experiments at Reinickendorf saw a flurry of activity. On the first anniversary of the Raketenflugplatz, a newsreel company came to film the launching of the one-stick Repulsor with the latest motor design. The rocket launch started out well but ended in disaster. After climbing to over 4,000 feet, the recovery parachute deployed but tore away and the rocket fell on top of a barn belonging to the local police department, some 3,000 feet across the road. Remaining hot fuel from the motor ignited the roof, and there was a small fire to deal with. It was shortly thereafter that the local police chief banned all rocket flights going forward. However, by the middle of October Nebel had convinced the authorities to let the experiments continue under slightly tighter safety regulations. Soon the testing continued, and by the end of the first year, the group had launched more than 80 rockets and conducted over 250 static firings of varying motor designs. Conspicuous at many of these tests was a fair-haired youth who seemed to be in the middle of every discussion.

Wernher Magnus Maximilian von Braun was born to Baron Magnus von Braun and Emmy von Quistorp on March 23, 1912, in Wirsitz, a town in the eastern German province of Posen. Wernher's father was a wealthy farmer and a provincial councilor and served as Minister for Agriculture during the 1930s in President Hindenburg's Weimar Republic. From childhood, Wernher revealed an interest in both science and music. At age 11 he enrolled in the Französisches Gymnasium that had been established two centuries earlier by Fredrick the Great. There, the boy showed only a modest ability in mathematics and physics, subjects in which he would later excel. In 1928 Wernher's father placed him in the progressive Hermann Lietz schools. Wernher's grades and abilities improved. Oberth's book captured the young boy's attention. However, von Braun soon learned that he would have to excel in mathematics to even understand the concepts and principles in the book. Even during these younger years of his life, von Braun was experimenting with rockets and propulsion. He once strapped a cluster of solid rocket motors to a wagon and shot it down a crowded street. Many in the crowd were not amused.

"I was ecstatic," von Braun later recalled. "The wagon was wholly out of control and trailing a comet's tail of fire, but my rockets were performing beyond my wildest dreams." The fire-breathing wagon diverged onto the Tiergarten Strasse, a very crowded Berlin city street. An angry police officer grabbed the young rabble-rouser and threatened to arrest him. "Fortunately, no one had been injured, so I was released in charge of my father."[5]

A pivotal point occurred for the then 18-year-old von Braun when he entered the Technische Hochschule in the Berlin district of Charlottenburg. While in Berlin, von

Braun's interest in astronomy and space travel continued to grow. He had become acquainted with Hermann Oberth, writer and spaceflight promoter Willy Ley, and rocket experimenters Rudolf Nebel and Johannes Winkler. He also followed the solid-fuel exploits of Max Valier. Von Braun quickly joined the VfR and was soon participating in rocket experiments in Berlin.

It was in Charlottenburg that von Braun studied under Professor Doktor Karl Emil Becker, a friend of his father. Becker was also a Lieutenant Colonel in the Reichswehr (German Army). As head of the Ballistics and Munitions Branch of the Army Weapons Department, Becker had long been involved with research and development of long-range artillery. During the First World War, he assisted in the creation of the Paris Gun and believed strongly in the importance of innovative new weapons development.[6]

German forces deployed the Paris Gun (German Kaiser Wilhelm Geschutz' long-range gun) in the later stages of WWI. It was commonly called the Paris Gun because of its use to bombard Paris from March to August of 1918. The gun was positioned on railway mountings 77 miles from the city of Paris. The 21-centimeter gun was manufactured using 38-centimeter naval guns fitted with special 40-meter-long inserted barrels. The shells weighed 265 pounds and were fired by a 400-pound powder charge, giving them a range of up to 81 miles. The humongous nature of such a weapon limited its mobility and strategic usefulness. Long-range artillery guns soon reached the point of being so massive that they were becoming impractical.

Colonel Becker's focus included liquid-fueled rockets. One year earlier, Becker had hired a young German Army Captain, fresh out of Charlottenburg with a master's degree in mechanical engineering, named Walter Dornberger. Captain Dornberger joined Becker's assistant, Captain Ritter von Horstig, along with Captain Leo Zanssen to form the nucleus of the fledgling German Army Rocket Program.

Many in the German Army believed that the First World War had never ended. Germany had not surrendered; the war was just on hold. German generals fully expected the cease-fire to eventually collapse and the fighting to recommence. The Allies anticipated this mindset. Therefore, the Treaty of Versailles placed severe restrictions on Germany's military strength. It limited the overall size of the German Army along with the total number and types of weapons it could maintain. During the postwar years, German rearmament was greatly hampered. Preceding Hitler's rise to power, the German military tried to operate within the framework of the restrictions, taking advantage of any omissions unforeseen after the end of WWI. This fostered new research into innovative weapons technologies such as rockets. The advancements in amateur rocketry of the 1920s caught the eye of several key individuals involved in German military arms research.

However, skirting the Versailles Treaty was not the primary reason for rocket research, especially later on, after Hitler began violating its terms incessantly. Rockets were seen as potentially superior weapons to artillery, having a longer range and greater mobility. This was particularly true for liquid-fuel rockets because they offered much greater range with heavier payloads than solid-fuel rocket motors. The development of the rocket had its roots in the enthusiastic amateur German rocket societies, which cultivated emerging specialists such as Wernher von Braun; however, it was the German military, using the emerging technology as a weapon for war, which shaped the V-2.

The Ballistics and Munitions Branch was solely interested in collecting real scientific data on rocket propulsion. Becker was not opposed to providing funding to private individuals or organizations if said individuals could produce usable data. The problem was usually that these groups drew a huge amount of publicity, something the Army wanted to avoid at all costs. The few joint projects entered into with private researchers produced very little success and worse, very little hard data.[7]

The VfR members may not have known it at the time, but soon their existence would come to an end. Late in 1931, one of the society's main financial backers withdrew funding from the VfR and redirected the money toward the experiments of Johannes Winkler. Although he was an established member of the VfR, Winkler had continuously conducted his own independent experiments. Earlier that year while the VfR was actively preparing at Raketenflugplatz, Berlin, Winkler had launched the first liquid-propellant rocket in Europe at another location. This produced some hard feelings within the VfR, whose members assumed that this honor would eventually become theirs in Berlin. The coming winter saw worsening economic conditions, which also contributed to the slow dissolution of the VfR membership. Increasingly, members were saying that they could not afford the club dues of eight marks. At the beginning of 1932, membership dropped to approximately 300.[8]

As the activities of the VfR diminished, in desperation, Rudolf Nebel wrote a report touting the benefits of using long-range rockets as artillery. He sought out Colonel Becker and gave him a copy of the report, hoping to secure German Army funding for VfR activities. Becker was willing to listen to Nebel's proposal, since the Army was making very little progress on its own. A few days later, Becker, along with Dornberger and von Horstig, traveled to the Raketenflugplatz at Reinickendorf to inspect the facilities. They were extremely disappointed. The rockets they were shown seemed very small and elementary. When Becker asked to be shown collected data such as thrust curves, fuel consumption, and internal temperatures, none could be given. In truth, if the VfR membership had one overriding deficiency, it was their lack of accumulating hard data. On April 23, 1932, the Army visited the Raketenflugplatz again and gave Nebel a small contract for 1,367 marks if he could build a rocket that would successfully reach 3,000 meters in altitude while ejecting a red flare to be tracked with Army instruments.[9] The launch would take place on a date in the near future to be specified by the Wehrmacht (German Army) at Versuchsstelle West (Experimental Station West), the new Army proving grounds at Kummersdorf.

The rocketry proving grounds at Kummersdorf had come about through the efforts of the Ballistics and Munitions Branch in late 1930. Becker picked a location on Army property because they wanted a professional facility where they could finally begin serious testing and start documenting the parameters of rocket engines. The Army facility at Kummersdorf could provide the logistics and security they needed. The necessary funds were procured through the Army Weapons Office, and in early 1931 work began at the Kummersdorf artillery range. Soon a test stand for solid-fueled rocket motors was erected, followed by installation of the latest measuring equipment that could be found.

It was a sunny July morning in 1932 when a handful of VfR members, including von Braun, loaded into their cars and drove south out of Berlin. They arrived near Kummersdorf, where they met Captain Dornberger at a designated rendezvous point. Dornberger led the group to an isolated location on the artillery range. The group was surprised to see numerous scientific measuring instruments already in place at the location, some of which were unknown to the amateur rocketeers. The VfR rocket was in place and fueled by mid-afternoon. At ignition the rocket vaulted a few hundred feet into the air, then it abruptly veered horizontally as it became unstable. It crashed nearby before the parachute could deploy. Disgusted with the pathetic spectacle, Becker refused to pay Nebel the agreed-upon price, saying the rocket's performance in no way met the requirements stipulated for the test. With the establishment of Kummersdorf, the Army now decided to cut all ties with Nebel and the VfR. It was time to stop the juvenile games and reconnoiter serious ballistic weapons research.[10]

By December of 1932, the Experimental Station West at Kummersdorf was growing. New buildings such as workshops, offices, drafting rooms, darkrooms, and a measurement room were constructed. In addition to the existing solid-fueled engine test stand, a new liquid-fueled engine test stand was added—the first ever established in Germany. Plans were finalized for their first designs and tests. For the next several months everyone on Dornberger's Section 1 team was either busy designing or constructing the components for their first rocket engine tests.

By 1932, political circumstances in Germany were in chaos, even worse than just one decade before, because of the worldwide economic depression following the crash of the American stock market in 1929. The Nazi Party almost won the presidency under a radical new revolutionary leader named Adolf Hitler. Only a year later, Hitler would be appointed chancellor of the German nation, and he would quickly seize full dictatorial powers, pronouncing himself Führer of the German people. His powerful words struck a cord that the German public wanted to hear. Hitler promised that Germany would regain its world status and power. German prosperity would rebound. However, he also stirred the innermost prejudices and hatred within the German society. Bigotry directed against minorities was encouraged, even fostered by the state, especially against Jews. This penchant would eventually figure prominently in the story of the V-2 rocket.

Hitler promptly crushed any potential political opposition. He formed around himself a circle of criminals and thugs who used assassination and intimidation to increase their stranglehold on power. The Schutzstaffel, otherwise known as the SS, became an army of personal bodyguards for Hitler. A separate entity from the German Army, the SS recruited the best of the best—the most elite soldiers with a sworn allegiance to Hitler. The ranks of the SS included some of the most ruthless and ardent Nazis. Heinrich Himmler was named head of this organization, which eventually carried out some of Hitler's most reprehensible proclamations.

Meanwhile, the VfR was dying a slow death. With its membership dwindling, the organization could hardly afford any sort of meaningful experimentation. Some of its most gifted members had moved on; a few were now employed at Kummersdorf. The end came commiseratively when a group of Nazi pilots began demanding access to the Raketenflugplatz for use as a training ground. The VfR refused them. Mysteriously, an outrageous water bill appeared from the city for over 1,000 marks, an amount the rocketeers could not afford to pay. The City of Berlin invalidated the VfR's contract, and the Nazi pilots gained for themselves a new training field at Reinickendorf—just like that.[11]

For amateur rocketry enthusiasts outside the realm of the German military or German companies, things were about to get tough. The perceived need for utmost secrecy and the desire to garner the most ingenious minds for a new military weapon generated a climate whereby any discussion or research from the outside had to be commandeered or suppressed. Even after Hitler gained power, conflicts and struggles for control within the National Socialist leadership continued. Having ruthlessly secured power for their leader, the SA, or Sturm Abteilung (storm troopers)—the Brownshirts—were abandoned by Hitler. The military and the SS convinced him of an impending coup by the SA, and what followed on June 29, 1934, was the infamous "Night of the Long Knives." During the next 24 hours, more than 70 senior SA officers were shot, including the head of the SA, Ernst Roehm. Very soon, rocket experimenters who happened to be associated with factions such as the SA were swallowed up or swept aside by the military.

Wernher von Braun had received his bachelor's degree in aeronautical engineering from the Charlottenburg Institute of Technology in the spring of 1932. In between his studies von Braun had found time to write an article entitled "Das Geheimnis der Flüssigkeitrakete" ("The Secret of the Liquid-Fueled Rocket"). It was published in the June 4, 1932, issue of the science magazine *Umschau*.[12]

Dornberger later wrote about his first encounters with the young, vivacious von Braun, "I had been struck during my visits to Reinickendorf by the energy and shrewdness with which this tall, fair, young student with a broad, massive chin went to work, and by his astonishing theoretical knowledge. It had seemed to me that he grasped the problems and his chief concern was to lay bare the difficulties. When General Becker later decided to approve our Army establishment for liquid-propellant rockets, I had put Wernher von Braun first on my list of proposed technical assistants."[13]

On November 1, 1932, von Braun signed a contract with the Reichswehr to conduct research leading to the development of rockets as military weapons. In this capacity, he would work for Captain Dornberger. In the same year, under a Wehrmacht grant, von Braun enrolled at the Friedrich-Wilhelm Universität from which he graduated two years later with a Ph.D. in physics. His dissertation dealt with the theoretical and practical problems of liquid-propellant rocket engines. Dornberger also began to recruit other VfR standouts such as Heinrich Grünow, an exceptional mechanic; Arthur Rudolph, a former colleague of Max Valier and engine designer; and Walter

Riedel, an accomplished researcher previously employed by the Heylandt Company.

The program at Kummersdorf started with some of the same dilemmas facing the VfR; namely, funding. The economic situation of the early thirties meant things were tight, even for the German military. Zanssen and Dornberger were having difficulty in securing even the most scanty supplies for the operation at Kummersdorf. The military budgetary bureaucrats scrutinized every purchase made by the facility. In the beginning they were not allowed to order such materials as machine tools or office equipment. These types of items were procured by using misleading descriptions on invoices or by simply classifying the purchase as secret. In 1933 Colonel Becker was promoted to chief of the Heeres-Waffenamt Prüfwesen of the Army Weapons Office. He was now in charge of allocating funds to various testing branches. This brought in a bit more money to Kummersdorf, but the funds were still limited.

Very slowly the operation grew in size and savoir-faire. Preparations were underway to construct a rocket that would finally take flight. Up to this point, only exiguous experimentation had been conducted. Fuel mixture, flow, cooling, and ignition had been studied, but only in static test conditions. A new rocket would be proposed using the moniker Aggregat (assembly) number 1 (A-1). Inherent in the proposed design of this rocket was the idea that the rocket or a portion of the rocket should spin to maintain stability. The rocket would stand 55 inches tall and be one foot in diameter with a weight of just around 330 pounds. Different from previous designs, the Heylandt-produced rocket engine was to be located at the bottom of the rocket, contained inside a portion of the alcohol fuel tank. At the top of this tank was an insert to accommodate a container for the liquid oxygen. Nitrogen was used to pressure feed the propellants to the engine. The overall weight of the A-1 came in at almost 400 pounds, and the spinning flywheel in the nose caused instability. It was not going to be a device that could fly, but it did provide valuable information about fuel mixtures and cooling. Fuel and oxygen-valve inconsistencies caused delayed and explosive ignitions. Three examples of the A-1 were built and test fired.

Next came a redesign of the basic A-1, renamed the A-2. This prototype maintained the same proportions and performance, but the stabilization device was moved to the center of the rocket. The 300-kilogram thrust engine was retained, but a separate liquid oxygen tank was added to prevent an explosion from mixing during powered flight. After preliminary tests were conducted, the team at Section 1 decided to test launch two prototypes of the A-2 design. The range at Kummersdorf was too small to conduct these tests in secrecy, so in December 1934, the two rockets, nicknamed Max and Moritz, were transported to the North Sea island of Borkum. The winter weather was somewhat forbidding; however, the group managed to successfully launch the first rocket on December 19, 1934. Climbing to just over one mile in altitude, the rocket fell onto the beach not too far from the launch tower. The following day the second example was launched. The 300-kilogram thrust engine burned for 16 seconds, and the rocket attained about the same height as the first. The rocket team was ecstatic. Here was a real rocket that had performed up to their expectations. Word of the success was sent to Dornberger at Königsbrück, who was on duty as commander with the first Nebelwerfer solid-rocket artillery batteries. Dornberger was pleased. They now had something to show for the investment made by the Army.[14]

In mid-January 1935, von Braun made a presentation to the leaders of Army Ordnance touting the progress of the rocket team in Section 1. The A-2 success brought new opportunities. A few days later, Kummersdorf received a visit from Major Wolfram von Richthofen, a relative of the famous Red Baron of WWI fame. Von Richthofen was the head of aircraft research for the German Luftwaffe. He was interested in developing rocket-powered aircraft, as well as jet-assisted launching pods for Luftwaffe heavy bombers. He asked the Kummersdorf team if they could design such systems. Working as a contractor, the Kummersdorf staff accepted the challenge—mainly because the Luftwaffe provided more research funds. A contract was signed, and within a few weeks, the Heinkel aircraft company brought their own engineers to Kummersdorf, helping to install a 1,000-kilogram thrust rocket engine in a Heinkel He 112 fighter aircraft.

An early test involving aircraft mated with rockets was the static firing of a 300-kilogram thrust rocket engine attached to an old Junkers A-50 Junior. With von Braun in the cockpit, the stationary airplane shook tremendously when the rocket engine was ignited. This old aircraft could not stand the stresses imposed by such a powerful engine.[15]

In early March of 1937, the modified He 112 was ready for its first test flight. Pilot Erich Warsitz started the airplane's Jumo engine, then readied the rocket motor in the tail. At ignition there was a massive explosion. The whole aircraft was blown to bits, but by some miracle, the pilot was unharmed. In April Heinkel donated a newer He 112 B for another test. This time the powered flight was successful.

Elsewhere in Germany, others were already developing jet-assisted takeoff mechanisms outside of Kummersdorf. Hellmuth Walter, working at the Germania-Werft, was developing a wakeless torpedo. He used hydrogen peroxide to fuel the torpedo motor. The hydrogen peroxide would condense quickly in water as it was expelled in the form of superheated steam, thus leaving no wake. This experimentation led to the development of a small rocket engine that was used as a booster to shorten the takeoff distances of various aircraft. Eventually, Walter would develop the HWK RI rocket motor, fueled with T-Stoff (oxyquinoline or phosphate), hydrogen peroxide, and Z-Stoff (calcium permanganate). This generation of aircraft rocket engines would later power the remarkable Me-163 "Komet," a rocket-powered, delta-winged fighter aircraft.

All during this time, Section 1 at Kummersdorf was developing larger and more powerful rocket engines. Showing great promise, these new engines provided thrusts of 1,000 and 1,500 kilograms. Several new test stands were constructed to accommodate these larger engine designs. The most advanced of these was a test stand designated for the prototype Aggregat 3 (A-3). The purpose of the A-3 was to conduct further tests with larger, more powerful rocket engines and to incorporate initial tests in fledgling guidance systems. As work began on the A-3, many other important advancements were occurring, not necessarily related to building rockets.

By this time in the United States, Dr. Goddard had developed and launched new rockets of ever-increasing efficiency. From 1930 to 1935, he developed rudder systems for steering a rocket in flight by deflecting the exhaust and gyroscopes to keep the rocket on course. However, the U.S. military showed no interest in his work. This lack of interest, along with some scurrilous newspaper articles during his early years of research, caused Goddard to become suspicious and secretive about his experiments. Even so, when he received a few unsolicited letters from German engineers asking technical questions now and then, he casually responded.

By 1936, it had become clear to most everyone at Kummersdorf that the seemingly small confines of Experimental Station West

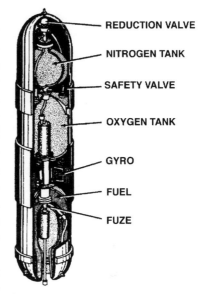

REDUCTION VALVE

NITROGEN TANK

SAFETY VALVE

OXYGEN TANK

GYRO

FUEL

FUZE

The A-2 rocket.

were unsuitable for test flights. The Kummersdorf range was not only too small for launching liquid-fueled rockets, it could no longer be expanded. In addition, security could be easily compromised if the populace of Berlin looked to their south and witnessed test missiles soaring skyward. Also, things were crowded. Section 1 workshops and facilities were crammed with over 80 people by this time. Dornberger eventually persuaded Major General Werner von Fritsch, head of the Reichswehr, to visit Kummersdorf in March of 1936. He wanted to convince General Fritsch of the deficiencies in their current situation and ask for a more secure environment in which to conduct the secret rocket experiments. Upon his arrival, the general was briefed about the ongoing research and the Kummersdorf inadequacies. Later, the general was escorted to the three captive test stands where he witnessed in succession the test firing of several engines. The impression must have been dramatic, because when the show was over, von Fritsch simply stated, "How much do you want?"[16]

The Army began contemplating the possibility of a large research center, a center that would be unique. It should be devoted to the development of a weapon unlike any seen before. Dornberger set the standards for selecting the new proving ground. It must be located on the coast near the water. The firing trajectory should be equidistant to a coastline for tracking purposes. The location should be flat and large enough for an airfield. Lastly, the center should be constructed in a remote location, away from view for the utmost secrecy and security. Von Braun had been conducting a search on his own initiative for the past several months all along the Baltic coast. A location on the island of Rügen was at first thought to be suitable, but there was no way it could be wrestled from the German Labor Front, as it was destined to be the official Nazi beach resort for all union workers. While visiting his parents, von Braun's mother suggested he look at Peenemünde. She said her father used to go duck hunting there. Von

Braun followed her advice and took a trip to see the area himself. It was perfect. The location met all of the requirements set forth for the new research center.[17]

Located on the northern tip of the island Usedom, Peenemünde was a small fishing village on the east bank of the River Peene. It was so small it was not even listed on many German maps. The nearest large cities were Rostock and Stettin, approximately 100 kilometers away. Having only one bridge connecting the island to the mainland, the site was relatively inaccessible. The coastline of the peninsula presented an uninterrupted stretch of German-controlled water 180 kilometers in length to the east. The small island of Greifswalder Oie was conveniently located just a short distance from the main body of Peenemünde, which would provide a location for remote tracking and secure testing . The area designated for the construction was then a nature reserve—a wilderness of thick trees, marshes, and dunes, relatively unspoiled by humans. Even today the dense forests of oaks and pines abound with wildlife.

The most important benefit from the rocket group's association with the German Air Force was the enthusiasm shown by the Luftwaffe officials. The Air Ministry was keen to expedite the development of rocket-powered aircraft. In a meeting with von Braun, the Luftwaffe's von Richthofen nonchalantly offered five million reichsmarks to the Army's rocket group toward the construction of the new facility. When word of this reached Army Ordnance, it caused an uproar. The unprecedented breech of military etiquette angered General Becker. Vowing not to be upstaged, Becker told Dornberger the Army would not be outspent by the "junior" service and pledged six million reichsmarks to the project.[18]

In the spring of 1936, a meeting was arranged at the Luftwaffe headquarters of Major General Albert Kesselring to work out the details of the new establishment. Present were Becker, Dornberger, von Braun, and von Richthofen. Kesselring liked the idea of the joint project and immediately approved the funds to purchase the property from the city of Wolgast. The northern tip of Usedom was to be divided between the Luftwaffe's Peenemünde West and the Army's Peenemünde East. The construction project was given to the Luftwaffe engineers. The rocket group appreciated the energy of this new, nonbureaucratic service. Many believed the project would move faster and more efficiently if carried out by the Luftwaffe's Air Ministry. Soon construction started.

It would take a few years to complete the construction at Peenemünde. In the meantime, work continued at Kummersdorf. The A-3 was still a priority, but now, assured of an establishment of such grandiose scale in the near future, the engineers turned their attention to a much larger project: the A-4, later to be known as the infamous V-2. It was time to start thinking of a weapon. The group came up with initial designs on paper, not very specific, but it was to be something big. The specifications for the A-4 were compulsory for the creation of an artillery round, not a spaceship. To make the A-4 look attractive to the Army, Dornberger decided the A-4 should have twice the range of the Paris Gun of the First World War. It should have a one-ton warhead and be easily transported on existing German infrastructure. The thrust required to propel a missile of this size would be about 25 metric tons. Thinking as a military

Dr. Robert H. Goddard's rocket after launch in New Mexico on April 19, 1932. In 1930, with a grant from the Guggenheim Foundation, Goddard and his crew moved from Massachusetts to Roswell, New Mexico, to conduct research and perform test flights away from the public eye. This rocket was one of many that he launched in Roswell from 1930 to 1932 and from 1934 to 1941. Dr. Goddard has been recognized as the father of American rocketry and as one of the pioneers in the theoretical exploration of space. (*NASA*)

man and artillerist, not as a space visionary, Dornberger prescribed a weapon intended to surprise and demoralize an unsuspecting enemy. He was very restrictive in his demands relating to accuracy requirements: for every 1,000 meters in range, a deviation of only two or three meters was acceptable. At a range of 230 to 250 kilometers, this would mean the A-4 should impact no further than 750 meters from the intended target. It would be much harder to achieve this than maybe he realized at the time. However, if it could be accomplished, the weapon would be quite formidable. If research continued as planned, it was hoped to bring the new weapon to the battlefield as early as 1943.[19]

The speed and efficiency of the Air Force construction at Peenemünde was nothing short of astonishing. In one year the project had progressed enough to allow most rocket activities to move north from Kummersdorf. The Army installations, called Heeresversuchsanstalt Peenemünde (Army Research Center Peenemünde), were much larger than the installations for the Luftwaffe. Peenemünde West and its adjoining airfield encompassed only about ten square kilometers on the western tip of Usedom. In May of 1937, the East Works of the Peenemünde Experimental Center opened with the young Wernher von Braun appointed as technical director. He had several hundred employees at his disposal; many nonprofessionals supplemented the group of scientists and engineers. The research establishment was spread out generally along the eastern shore of the peninsula, almost 15 kilometers north to south. North of Karlshagen, the planners built a new housing settlement for over 3,000 future employees. In addition to the neat row houses in this model village, there were shops, schools, a sports field, and a club. North of this settlement was the East Works headquarters and administrative buildings, laboratories, and workshop areas. A huge coal-fired, steam-turbine power plant generating 30,000 kilowatts of electricity for the complex was constructed on the west side of the peninsula, close to the Peene River. Along the eastern shore, between the sandy beaches and the established tree line, nine first-class test stands would be built. The most important of these would be known as Prüfstand VII (Test Stand VII) and was designed to be the ultimate testing and launching facility for the big A-4 missile. The whole peninsula was linked by a railway system. In just a few short years, the multitude of workers transformed the dense wilderness into one of the most advanced research centers in the world.[20]

The rocket group became an independent section headed by Dornberger, and they gained a new title—Wa Prüf 11 (Ordnance Test 11)—which was a part of Army Ordnance's Development and Testing Division. The new large test stands at Peenemünde would not be fully completed until 1939; however, engine development continued at Kummersdorf under the supervision of Dr. Walter Thiel. Thiel was born in Breslau. He studied materials engineering there at the technical university and began researching the use of hypergolic fuels for rocket propulsion. He joined Wernher von Braun's team at Kummersdorf in 1936, replacing Dr. Kurt Wahmke, who was killed in an explosion in 1934. Thiel was brought in purposely to develop larger, more powerful rocket engines. The bulk of the engine development group would remain at Kummersdorf until 1940, although Thiel would work closely with Klaus Riedel in setting up the test stands at Peenemünde.[21]

After the A-2 success in late 1934, von Braun planned the A-3, a larger and heavier rocket. The A-3 designers adopted the validity of many A-2 components. The 1,500-kilogram thrust engine in the A-3 was simply a scaled-up version of the A-2 power plant. Slight variations were made, a double-wall cooling method was introduced whereby the alcohol circulated around the combustion chamber, and the injection system utilized a different method of mixing the fuels, which created more efficient combustion and higher exhaust velocities. Along with more powerful engines, the difficult problem of guidance needed to be addressed to achieve the goals outlined for the future A-4. The expertise needed to manufacture a three-dimensional gyroscope was beyond the capabilities of the Kummersdorf staff in the early thirties. A stabilization device of this kind would be required for guidance and control of the A-4 to realize the accuracy requirements stipulated by Dornberger. An outside company specializing in naval gyroscopic manufacturing—Kreiselgeräte GmbH (Gyro Devices, Ltd.)—was contacted by the rocket team in hopes that it could produce a device to meet the needs of rocket guidance.[22]

By late 1936, after several design revisions, the device was delivered for installation in the A-3. The gyro platform would send signals to jet vanes situated directly in line with the rocket's exhaust, which would deflect the aspect of thrust. Connectors that extended from servos in the control compartment controlled the vanes. The A-3 was also equipped with long but narrow fins for aerodynamic stability. The idea was to design a fin with enough surface area to maintain the center of pressure behind the center of gravity but at the same time not present a heavy drag inhibiting the speed of the rocket. The shape of each fin would also need to maintain stability at supersonic velocities.

As early as January 1936, the A-3 design had been undergoing wind-tunnel tests under the supervision of Dr. Rudolf Hermann at the Technical University at Aachen. The tunnel used for the tests was extremely small, but the data indicated some disturbing issues concerning the overall A-3 configuration. It revealed that the A-3 was stable but so stable that it was susceptible to crosswinds that would cause flight deviation. Also, the fins did not provide a large enough profile to control the rocket at high altitudes and were going to burn up in the rocket exhaust, which would expand as the air density decreased. However, because it had taken more than six months to gather the information, the A-3 was more or less complete by the time von Braun received the results. It was apparent to Dornberger and von Braun that a larger, more sophisticated wind tunnel would need to be constructed at Peenemünde.

The first A-3 was ready for launch at the end of 1937. It had been an exciting year for the group. Having moved into the partially completed facilities at Peenemünde in the spring of 1937, they finalized the assembly of four A-3 prototypes and now were ready to launch from their new facility. However, the test flights were actually conducted on the tiny island of Greifswalder Oie, just a short distance from the tip of Peenemünde. Earlier that year, crews constructed a concrete launch platform along with an underground observation bunker, near the edge of the tree line. Control cables ran from the platform to the bunker and a telephone line was connected to the lone lighthouse nearby.

The weather was abysmal. Conditions on the Greifswalder Oie were the worst imaginable. Rain, wind, and cold delayed the launches. It was not the most ideal setting to conduct important rocketry experiments. The fact that the trials went on in these horrible conditions was evidence of how important the pace of rocket development was; the feeling existed that schedules must be adhered to, come rain or shine. In spite of the conditions, excitement and camaraderie among the crew kept spirits high as they prepared for the tests. The first A-3 was launched on the morning of December 4, 1937. The liftoff was good, then unexpectedly, the parachute deployed prematurely. The rocket turned into the wind and crashed some 300 meters from the launch site. The early parachute deployment caused a misdiagnosis of the flight deviation, and this was only confounded when it happened again a few days later during the second A-3 test flight. The parachute was removed during the third launch on December 8, but still the rocket turned into the heavy winds and crashed. A fourth launch attempt yielded the same results. Dornberger and von Braun remembered the predictions of Dr. Hermann; the A-3 wind-tunnel tests were proven correct. The rocket was susceptible to high winds, and the control system was not adequate to make the necessary adjustments to correct the course.

A new test missile was needed to iron out the steering problems still facing the rocket team at the conclusion of the A-3 tests. The wind-tunnel tests at Aachen and subsequent suggestions for improving the overall A-3 design convinced Dornberger to pressure Dr. Hermann to join the experimental staff at Peenemünde. In April 1937 Dr. Hermann was persuaded. The world's most sophisticated supersonic wind tunnel, which would ultimately simulate a running speed of over Mach 4, was built in the heart of the laboratory and workshop area of Peenemünde East.

After joining Peenemünde, Dr. Hermann quickly recruited the best aerodynamicists he could find. One of the first to join Rudolf Hermann in Peenemünde was Dr. Hermann Kurzweg. It was Dr. Kurzweg, working with hand-carved models with various fin configurations, who developed the basic design of the A-5 and later the A-4. The designation A-4 was already given to the final production version of a weapon, so the new test-bed missile was given the number A-5, even though it was out of sequence. It would resemble a miniature A-4, incorporating the same aerodynamic design as the future weapon along with a new inertial guidance control system. Internally, many of the A-3 components remained the same, and the engine was powered by the same fuels. But the A-5 would carry a radio-command system for ground control of engine cutoff and remote parachute deployment. While construction moved forward on the wind tunnel at Peenemünde, testing at Aachen continued. Small-scale models were launched to test the different fin designs, followed by the first actual test flights of the A-5 in October of 1938.[23]

Dornberger had dictated a regimented schedule for A-5 production and testing. The A-5 should have the ability to carry out all of the flight research tasks that were essential to final design requirements on the larger A-4. This included the ballistic shape, the ability to pass through the sound barrier, and guidance throughout the burning portion of the missile's flight. While waiting on industrial contractors to put the finishing touches on the new gyroscopic equipment, the Peenemünde team launched four unguided A-5s from Greifswalder Oie and was pleased with the results of each flight.

The political factions in Germany were now dominated by the Nazi Party. The rocket team, in their daily bustle, paid only slight attention to the international struggles taking place at the time. Events in Europe were quickly heading to a boiling point. Hitler gave notice to the military leaders in Germany, and to the world, when he began rearming Germany after 1935 that confrontation was inevitable. By tearing up the Treaty of Versailles, he instigated a calculated process of testing his European neighbors' resolve. The Franco-Soviet pact allowed Hitler to move into the Rhineland and reclaim German territory without repercussions. Japan, Italy, and Germany were allied in an axis, which worried many nations around the world. Austria, Bohemia, and Moravia soon fell under Nazi influence, with Hitler's real objective being the overrunning of Prussia and the lands to the east.

It was during this time that relations between the Nazi leadership and the heads Germany's armed forces clashed. The generals saw all too clearly the direction Hitler was steering the country. The friction and mistrust between the two factions eventually caused high-ranking Nazis to drum up fictitious charges against Army Chief von Fritsch. On January 25, 1938, Hitler was shown a document in which Reichsführer SS

Himmler accused General von Fritsch of criminal homosexual activities. Von Fritsch, creator of the modern German armed forces, strongly denied these accusations, but on February 4, 1938, Hitler had him removed. Von Fritsch was replaced by Colonel General Walther von Brauchitsch, a man considered by his peers to be one of Germany's best and brightest military leaders. Von Brauchitsch, a former artillerist, was seen as a man of culture, integrity, and charm. Although he was personally disinclined to Nazism, opposing Hitler's invasion of Austria and Czechoslovakia, he was also highly ambitious and keen to succeed von Fritsch as Commander in Chief of the Wehrmacht. Seeing the opportunity, Hitler abolished the War Ministry, reorganized the armed services, and created the Armed Forces High

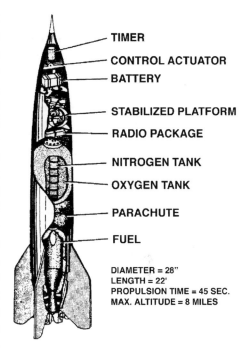

TIMER

CONTROL ACTUATOR

BATTERY

STABILIZED PLATFORM

RADIO PACKAGE

NITROGEN TANK

OXYGEN TANK

PARACHUTE

FUEL

DIAMETER = 28"
LENGTH = 22'
PROPULSION TIME = 45 SEC.
MAX. ALTITUDE = 8 MILES

The A-5 rocket.

Command (OKW). The Führer assumed full command. Von Brauchitsch understood Hitler's intentions. If Hitler was determined to take the country to war, von Brauchitsch would strive to make sure Germany's armed forces were ready.

It was a cold, rainy day in the spring of 1939 when Hitler visited the Kummersdorf Experimental Station with Field Marshal von Brauchitsch and General Karl Becker of Army Ordnance. Others in attendance included Deputy Führer Rudolf Hess, Martin Bormann, and several others. By this time, Peenemünde would have been more representative of contemporary rocket research, but because of the extreme secrecy surrounding the new rocket center, the Führer did not visit Usedom. After some introductions, Dornberger, now a colonel, proceeded to escort the entourage around the old facility. Dornberger described to Hitler the research at the station, providing a basic outline of the group's history and current objectives. To some in the tour, Hitler seemed to be somewhat disinterested. They followed Dornberger to the captive test stands, where in preparation for the Führer's visit, several engine tests had been readied. With cotton stuffed into his ears, Hitler peered through the observation slot of a protective wall as a 300-kilogram thrust engine was fired. Those accompanying Hitler were smiling and excited by the demonstration, while Hitler said nothing. Next was another static firing of a more powerful 1,000-kilogram thrust engine. The noise from the second engine made Hitler wince, but otherwise he showed no emotion whatsoever.[24]

As the party walked over to the third test stand, Dornberger briefed Hitler about the progress in Peenemünde. A model of the A-3 was laid out showing the various components, and at this point von Braun began explaining the inner workings of the rocket. Hitler became gradually more interested, inspecting the rocket and listening closely to von Braun's comments. He even asked a few questions, one about the need for such exotic fuels. Von Braun responded respectfully, explaining the need for high exhaust velocities to obtain the speeds necessary to extend the range of the rocket. At the conclusion of von Braun's speech, Hitler turned away shaking his head. It is unknown if Hitler discounted the feasibility of the rocket or if he was just overwhelmed by the complexity of it all. Later in the station mess, Hitler was served a light lunch and during the meal talked with General Becker about various details of what they had seen. When finished, Hitler simply said, "Es war doch gewaltig!" (It was nevertheless grand!)

Hitler's attitude toward the rocket at this time was certainly perplexing to Dornberger and others. Dornberger recalled later, "He was the only visitor who had ever listened to me without asking questions." For a man with such zeal when it came to new weapons like warships, guns, and tanks, Hitler's reluctance to embrace the innovative technology of the rocket was hard to understand. The Führer was always keen on the "next big project," so why is it he discounted the rocket? His ineptness was not limited to understanding the technology of the rocket, as he would make similar blunders all through the war, imposing his supposedly infallible will on many other innovative projects and thereby negating their usefulness.[25]

If Hitler was unimpressed by the demonstration at Kummersdorf, he changed his mind, if only briefly, a few months later. On September 19, 1939, at a rally in Danzig following the invasion of Poland, he spoke openly of a weapon "which is unknown to others, and with which we ourselves could not be attacked." It made for good propaganda; however, with his notions of weaponry still shaped around battles of centuries past, Hitler did not truly grasp the usefulness of the ballistic missile in modern warfare.[26]

In the fall of 1939, the Peenemünde Research Establishment was fully staffed after the transfer of the remaining personnel from Kummersdorf. The research center was under the control of Walter Dornberger, while Colonel Leo Zanssen was retained as military commander. For one to imagine the atmosphere, the logistics, and the sheer spectacle leading up to and during the test flights, it is crucial to have some comprehension of the layout of the facilities. The experimental test stands were carefully planned and constructed in the northern portion of the peninsula near the seashore. Each area was strategically located by function and sheltered by the dense Usedom forest. The largest test stand, Prüfstand VII, otherwise known as P-7, was the heart of the testing center and the location for most test flights. P-7 was the most advanced rocket-testing facility imaginable. It was surrounded by an oval earth wall rising 12 meters high to provide protection from the winds blowing off the sea. A short distance away, just outside the wall, stood the giant assembly building—a concept that was adopted 20 years later during the Apollo program at Cape Kennedy. The P-7 assembly building rose 100 feet high and was 166 feet wide by 200 feet in length. The preliminary preparation of each rocket was performed inside the building. When ready, the rockets were mated into one of the mobile test stands or towed by trailer into the center of the P-7 arena. The mobile test stands, built to handle engine tests of up to 200 tons of thrust, could be moved from the assembly building, around the outer rim of the earth wall, and into the P-7 arena by means of a rail system mated with a gargantuan motorized platform. P-7 also featured a water-filled trench to suppress the heat of engine test firings, along with a DC generator and batteries, a blockhouse for cameras, storage for cooling water, and alcohol storage. Large pipes were built inside the earth wall, and these delivered 130 gallons of water per second for cooling purposes during tests. At the southeast end of the wall was the control blockhouse. This was a hardened concrete square building protruding from the earth wall with six control rooms of various functions.

Although Prüfstand VII was the main hub for flight testing, there were other various test stands nearby. Prüfstand X was located on the northern tip of the peninsula and was another launching area for large-scale rockets. A large open area that was accessible to vehicles, P-10 could be used to simulate actual field operations. A few hundred feet to the south of P-7 was the tall concrete and steel structure of Prüfstand I. Rising 50 feet high just inland from the seashore, P-1 was built to handle the big engines with a capacity of up to 100 tons of thrust. It was heavily used during the development of the 18-cup injector system for the A-4 engine. Moving further south, down the eastern side of the peninsula, there were several more test stands. The first was Prüfstand VIII. Here the fully assembled A-4 received engine startup tests along with testing of onboard instrumentation and steering controls in the jet blast. Next were Prüfstands IX and II, used in testing the Wasserfall antiaircraft missile and the A-5 engine, respectively. These were followed by Prüfstands IV and III, which were testing areas for smaller engines or alternative fuels. A little further on was Prüfstand V, used for calibration of pumps, steam turbines, and steam generators. The last in line was Prüfstand VI, which was an exact replica of the big test stand at Kummersdorf and had seen much activity during the testing period for the A-5 rocket.

Situated off the main road among the buildings of the Entwicklungswerk (Development and Experimental Works), the Administration Building was a 10-minute drive to the south. Adorned with a swastika and framed by the concrete pillars of a neoclassical style entrance, this building housed not only the administrative offices but also departments such as the Technical Design Department, the Future Projects Department, and Wernher von Braun's office. Various laboratories were located nearby with smaller shops supporting the laboratories. The Teilewerkstatt (Parts Production Plant), the Aerodynamisches Institut, and the Mach 4 supersonic wind tunnel, along with the Guidance Laboratories, were all in proximity to one another. Further west across the peninsula were the liquid oxygen factory and the rocket center's power plant. The development workshops were known as Werk Nord (North Works). To the south there was dense forest for a little over one kilometer.

Driving south from the East Works, it was hard to miss the giant building looming ahead. The A-4 production facility, Building F1, was a modern assembly plant for rocket serial production. Supporting this building were several small workshops. Eventually, there would be a second production building built next to it of similar size and stature. This area was known as Werk Süd (South Works). South of this area was the housing estate for Peenemünde personnel and a German Army compound. The Army and Luftwaffe facilities covered almost the entire peninsula.

Although the facility was finally fully staffed, a new problem arose. Now that Germany was at war with France and Great Britain, manpower at Peenemünde was slowly being siphoned away due to increasing conscription of men for the German military. Dornberger, along with General Becker, met with Army Chief von Brauchitsch at Army headquarters in Zossen to discuss this problem. Von Brauchitsch was persuaded to sign an order stating that "Peenemünde should be pushed forward by all possible means." It was hoped that such a directive would alleviate the drain of important personnel and skilled workers. General Von Brauchitsch had been Dornberger's superior officer in the 1920s. He viewed Dornberger as his junior protégé and kept a close watch on his burgeoning career. Months earlier, in November of 1938, Dornberger had convinced von Brauchitsch to issue a directive for the construction of a full-scale missile assembly factory at Peenemünde under the auspices of it being "particularly urgent for national defense."[27]

Dornberger had no qualms about using his close relationship with von Brauchitsch to Peenemünde's advantage. Now that Germany was at war, funding for military endeavors would be quickly consumed by a myriad of new projects, and Peenemünde needed its share. Following Hitler's visit to Kummersdorf, Dornberger gained a greater understanding of the Nazi leadership's attitude toward new technologies. It seemed as long as the experimental station could show progress toward the feasibility of a new weapon, the funding would continue.

By the end of the decade, the German Army's liquid-fuel rocket program had come a long way. The once-tiny organization had grown beyond the wildest dreams of its early participants. Throughout the late 1930s, activities and personnel continued to grow at the Peenemünde complex. Reichsmarks were being spent at an astonishing rate. A-4 development forged ahead with crucial technological breakthroughs occurring at regular intervals. However, without an operational weapon to show for their efforts, the work was far from finished.

The most daunting challenge was the development of the inertial guidance platform. Realizing the old control system installed on the A-3 was inadequate and that the new gyro-autopilot design would not be available anytime soon, the engineers decided to install a heavier, more powerful control device manufactured by the firm of Siemens. The A-5 was under construction after the finalizing of the new tail surfaces, which were redesigned and shortened after extensive wind-tunnel tests. It was thought that with the new tail surfaces being much more streamlined, the speed of sound might be achieved during A-5 test flights. In 1938 several small A-5 models made of solid iron were released from a Heinkel He-111 flying at 20,000 feet. The drop tests were recorded by phototheodolites and revealed that at around 3,000 feet, the dummy A-5s exceeded the speed of sound.

By the summer of 1939, Peenemünde was ready as a launching station. But it was

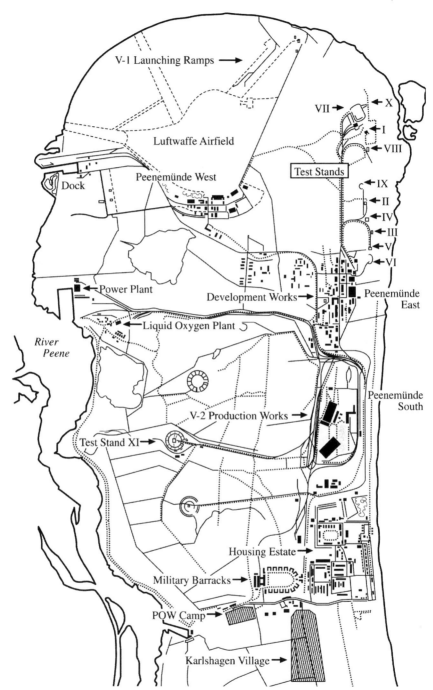

Peenemünde was situated on the Usedom peninsula along the Germany's northeastern Baltic coast, about 110 miles north of Berlin. Before the end of the war more the 30,000 persons had been assembled here, both voluntarily and involuntarily, to work on Hitler's most advanced weapons development.

on the tiny island of Greifswalder Oie that the rocket team set up for the A-5 experiments. Many changes had taken place on the island. Concrete roads, a concrete observation bunker, new sleeping quarters, and a large measurement house were now in place on the island. Special instruments were set up on the neighboring island of Rügen along with more instruments at Peenemünde. Each location was connected to Greifswalder Oie by cables laid under the sea. The closest station for measuring the trajectory of the rockets launched from Oie was located on the eastern end of Rügen, across ten kilometers of open water.

The first true example of the A-5 was ready for its maiden flight. The previous A-5 rockets launched in late 1938 carried no guidance systems and were launched only to test structural and aerodynamic changes from the A-3 technology. The A-5s would now be put through a regimented series of test flights intended to prove the viability of components and ideas destined to be integrated into the larger A-4 rocket. The weather conditions were good, with very little wind, as the first full test of the A-5 commenced from Greifswalder Oie in late October of 1939. At ignition the rocket climbed straight up, and just as planned, the exhaust vanes directed the A-5 on an easily controlled vertical flight course. The newly installed Siemens gyro-control gear was, so far, working as expected. When the motor shut down 45 seconds into the flight, the rocket was almost five miles high. Its momentum carried it higher until gravity slowed the ascent. At the high point of the trajectory, von Braun sent a radio signal to command the release of a drogue parachute, followed seconds later by another signal that released the main parachute. The rocket drifted slowly down, landing close to the island in the waters just off shore. The rocket was retrieved easily and taken for inspection.

The following day two more A-5s were scheduled for launch. The results of the morning flight produced almost the same results as those from the previous day. Once again, the rocket made a vertical ascent, straight up, without the complexity of an altered trajectory. However, Dornberger and the engineers were still cautious of celebration. Even though there was reason to be excited about the first two test flights, the A-5 was yet to perform its most crucial task.

The A-5 was a test-bed rocket for all principle features of the proposed A-4. Equipped with the new Siemens control equipment, it was designed to execute commanded guidance during a ballistic (curved) trajectory and have the ability to do so in stable flight. The rocket would maneuver into a ballistic attitude when the gyroscopes tilted in the desired direction of flight, which would cause the autopilot to send signals to the servos attached to the exhaust vanes. They, in turn, would deflect the blast in a manner so as to tilt the rocket slowly over. If wind gusts affected the attitude of the rocket, the autopilot would react in the same manner, always seeking to align the longitudinal axis of the rocket with the fundamental axis of the gyroscopes. Thus, the gyros were responsible for the controlled tilt during a curved flight path and primary flight course correction of the rocket. In laying out the requirements for the A-4, the scientists determined that a 50-degree tilt would be necessary to achieve the maximum range for the future weapon.

The third A-5 test flight, the second flight that day, was predetermined to test this technique of controlled guidance. The engineers had often tested the procedure during static firings of the rocket at the captive test stands, but now they would be able to see if it really worked in flight. As the rocket blasted away vertically from its launching point, it was only a few seconds later when, gradually, the programmed tilt came from the control system. They watched and cheered as the A-5 canted to the east after four seconds of vertical climb. It crossed over the island gaining speed as it flew in a long arc out to sea. When the motor stopped, the missile continued and flattened out about four miles downrange. Surprisingly, the parachute deployment was again successful, and the rocket dropped slowly from the sky into the waters of the Baltic. Once more, the rocket was recovered and subjected to post flight examination.

The A-5 guidance test was completely successful. Although the rocket had not achieved supersonic speeds, the calculations and devices worked as planned. In the A-5 the rocket team now had a proven tool with which all the varied concepts that would need to be incorporated in the A-4 could be tested. Later on, the A-5 would achieve a range of around 11 miles at a height of 8 miles. Dornberger was relieved. He later stated, "Now I can see our goal clearly, and the way that lead to it. Then I knew we would succeed in creating a weapon with far greater range than artillery." The A-5 would be launched again and again to test these concepts as the team moved closer to creating the big missile.[28]

The research facilities near the Baltic were not as confidential as the Army might have liked to believe at the time. The recent Peenemünde successes were, in part, the result of cooperation with civilian firms and German universities, all of which were privy to some form of confidential information about the rocket project. The first warning about Germany's ongoing secret weapons research was delivered to the British as a gift.

On the morning of November 5, 1939, a package was found resting on a window ledge outside of the British Embassy in Oslo, Norway. The package contained seven pages of German text and another small box, which contained a sealed glass tube. When the text was translated it sounded incredible. The document spoke of fantastic new weaponry being developed in Germany: remote-controlled gliding bombs, huge rockets stabilized using gyroscopes, advanced radar systems, and new types of torpedoes. The document was signed, "From a German scientist who wishes you well." Late in the evening of the same day, the so-called Oslo Report arrived on the desk of Dr. Reginald Victor Jones, the director of the Scientific Department at Air Ministry in London. Dr. Jones scrutinized the documents and then cautiously opened the small box, which was found to contain a glass tube with some sort of electronic device inside, later determined to be an electronic switch for triggering a proximity fuse. He immediately sent the device to be analyzed and circulated copies of the report to the three service ministries. No one believed the information to be genuine. It was quickly denounced as a hoax, one designed to intentionally mislead British war planners. Dr. Jones was one of the few who actually retained his copy of the Oslo Report, and its value would become apparent at a later date.[29]

However, Peenemünde was years away from actually fielding a weapon. Much more research requiring additional funding would be needed before the operational deployment of any future weapon. During Dornberger's earlier visit with von Brauchitsch in 1939, he expressed the need for more funding to support Dr. Thiel with his innovative research on the large A-4 combustion chamber. With their latest accomplishments, Dornberger was confident they could begin series production of the A-4 by 1943, if not sooner. It all depended on continued funding for the research center, but resources in Germany were already stretched to the limit. Even though Dornberger was able to secure the support of Army Chief von Brauchitsch, who issued two Army directives—one for the construction of the assembly plant at Peenemünde and the other for priority in manpower and material—both directives were made without consulting either Hitler or the Armed Forces High Command (OKW). The incoherent priority system made it impossible to meet von Brauchitsch's demands. Not only was skilled labor almost nonexistent but the interference of OKW staff resulted in Hitler canceling the directives. The Führer dictated that rocket development should continue at the agreed-upon prewar levels.[30]

From a distance, Reichsführer SS Himmler kept a close eye on rocket activities in Peenemünde. Any installation, such as Peenemünde, receiving so much attention and funding was bound to be noticed by the head of the SS. It is apparent that Himmler was interested in spreading his influence within the burgeoning rocketry program as early as 1940. Wernher von Braun was contacted by Himmler on May 1 and the Reichsführer SS awarded von Braun the SS rank of Untersturmführer (lieutenant). Dornberger, always looking for support of the rocket program, suggested to his colleague that it would be unwise to decline the offer. Von Braun was hesitant but finally accepted this position so as not to offend Himmler during a time when the rocket program was struggling for priority.[31]

During the next few years, Peenemünde would increasingly find itself garnering the attention of Germany's leadership and industry. Ironically, it was something that Dornberger had longed for but also feared. Would it be possible for Peenemünde to harvest support and maintain the integrity of their research? It was a supposition that proved difficult to achieve.

2

Pennemünde Research

For the scientists and engineers working in the peaceful surroundings of the Usedom peninsula, the war was far away. As battles raged for Hitler's armies during 1939–1940, life at the research center went on as usual. The massive complex was full of activity. To this point the ambitious construction project had consumed some 550 million Reichsmarks. Heeresversuchsstelle Peenemünde, HVP, occupied the eastern part of the peninsula, while the Luftwaffe's Erprobungstelle Karlshagen (Karlshagen Experimental Station) occupied the western part.

After the invasion of Poland, Hitler turned his blitzkrieg to the west seeking more territorial gains. In April of 1940, German forces invaded Norway and began an occupation of Denmark. In May Germany's Panzer divisions slashed through Holland, Belgium, and France. The French Maginot Line proved to be completely useless as panzer generals such as Guderian and Rommel steamrolled to the coast. The British Expeditionary Force, trapped at Dunkirk, was scrambling across the English Channel by June. Hitler was elated. His fast-moving armored columns and ferocious air power combined to completely overwhelm enemy opposition.

Just when Dr. Thiel was realizing appreciable results in his work on the A-4 engine and the A-5 flight trials were well underway, the rocket team was dealt a harsh blow. Less than six months after von Brauchitsch had placed a high-priority order on the Peenemünde work, Adolf Hitler cancelled the directive. Claiming Germany would have no need for the enigmatic ballistic missile, he had convinced himself that existing weapons systems could do the job of waging war. Funding the improvement of existing weapons made more sense, since he believed the war would be over in a few months anyway. Increasingly, he saw the rocket as superfluous and believed it ridiculous to divert technical and material resources to an unproven enterprise such as the ballistic missile. Also at this time, Hitler was engaged in many command frays with the armed forces, and Peenemünde became an inadvertent victim of this struggle. Without an operational missile on hand, the Nazi leadership would pay no attention to the previous ten years of progress established under Dornberger.

V-2

The engineers at Peenemünde knew they must be diligent. Time was running short, and the pressure was on to come up with substantial evidence of the rocket's viability as a weapons system. Their work was made even more difficult when a number of experienced personnel were drafted into the armed services. In September of 1939, von Braun invited over 30 university professors to Peenemünde for a three-day conference. Seeking to garner the help of educated minds from around Germany, von Braun hoped to alleviate the difficulties caused by the shortage of manpower at Peenemünde. At the same time, the research universities would benefit from additional government funding. Some of Peenemünde's most challenging technical problems such as tracking, computing devices, accelerometers, transmitters, receivers, and gyroscopes would be shifted to these research institutions. It is remarkable that in less than two years, most of these problems were successfully addressed.[1]

In March of 1940, Hitler moved to centralize weapons development under the auspices of one man with the appointment of Dr. Fritz Todt to the position of Munitions Minister. Todt was a dedicated Nazi Party member and admirer of Hitler. He had proven his resourcefulness as Generalinspektor für das deutsche Straßenwesen (Inspector General for German Roads, the builder of the autobahn highway system) and the West Wall defenses. In 1938 he founded Organization Todt, joining together government firms, private companies, and the Reichsarbeitsdienst (national work service). Although loyal to Hitler and the National Socialist State, Todt was a modest man, thought by many to be very reliable and one who stayed clear of political intrigues. He was unlike others in the governing class of the regime, living a quiet and secluded life outside of his work. He somehow managed to maintain his personal independence in his dealings with the Führer.[2]

The individual military service branches, which before had initiated their own research programs in conjunction with German industry, now found themselves subject to the whims of political bureaucrats in Berlin. In the months prior to Todt's appointment, weapons production figures had been declining steadily, and this gave Hitler the impetus to wrestle even more control from the armed services. The problem was that this politically motivated manipulation of industry was not guided by any technical or administrative expertise. The motives of the Nazi leadership were far more malevolent than just improving production figures, and within a few years, party officials would completely subjugate munitions funding to Berlin.

The neglect to the rocket program resulting from Todt's appointment was to have far-reaching effects upon the continued research at Peenemünde. Todt saw the rocket program as a waste of valuable resources. The ministry ignored the requests made by the military, in particular Peenemünde, to include the representation of scientific minds in the decision-making processes. Only a few industrialists, those with strong Nazi Party ties, were represented. Dr. Todt was eager to display his allegiance to Hitler and Nazi Party doctrine. It may be that Todt's lack of interest in rocket research was spurred on by the complete indifference paid to Peenemünde by party officials. Moreover, Munitions Ministry officials were quick to point out that they saw very little progress for the amount of funding currently being directed toward rocket research, and many felt Peenemünde was not providing a positive return on their investment. This is understandable when one considers their inability to grasp the potential of innovative fledgling technologies. However, there was also a real need to boost Germany's contemporary arsenal. Military supplies in 1940 were not as abundant as the propaganda machine led the German people to believe.

Not too long after Todt's appointment, General Becker, head of the Army Weapons Department and patriarch of German rocket research, took his own life. Becker's despondence over the continual struggles with political leaders brought him to the brink from which he could not return. Believing his status to be greatly diminished, he committed suicide. He was replaced in his position by General Emil Leeb, who was, as it turned out, just as supportive of the rocket project. Leeb, also an artillerist, believed that the rocket could be used as a deterrent. This was a progressive idea at the time, as it was not foremost in the minds of Germany's top military or political leadership. In the fall of 1940, General Leeb, in attempt to speed up the construction of the A-4 production plant at Peenemünde, decided to remove the Army construction office from the job and replace it with Albert Speer's firm called Construction Group Schlempp, which was brought in at the end of September.[3]

At times the problems facing the Peenemünde technicians seemed overwhelming. The complex rocket motor presented the majority of the difficulties. Not only did the engineers face the dilemma of new technological barriers to be conquered, the team needed to deliver an engine design that could be transitioned for mass production. By March of 1940, there was a noticeable increase in activity around the tall girders of Test Stand I. Suspended in the tower, just off the seashore, was the first example of the big engine for the A-4. The engine had undergone several major design changes—the most important being a double-wall design for cooling. Four welded rings with small inlet holes would allow alcohol to flow around the combustion chamber. The chamber pressure would force the alcohol against the inside walls and act as a shield between the hot gases and the chamber walls. On March 21, 1940, the engine was test fired successfully. The deafening roar from the test stand echoed across the complex as 25 tons of thrust was achieved lasting for 60 seconds. Even with the successful firing, the design would need considerable refinements to overcome several continuing technical problems.

Never before had anyone attempted to design an engine of this magnitude. The lack of available experience resulted in a series of trial configurations in which Dr. Thiel tested a variety of sizes and shapes for the big combustion chamber. In the end, the big barrel-like combustion chamber would feature an injector system at the top with 18 separate orifices through which alcohol would be brought to mix with the liquid oxygen. It was Dr. Thiel who was responsible for the enhanced engine performance by increasing reliability and efficiency during the transition to the A-4. Thiel's greatest contribution was likely his breakthrough decision to use turbo pumps to feed fuel into the rocket engine. Engine designs prior to this had relied on pressurization of the fuel tanks to drive the fuel out when a valve was opened. The turbine itself would be

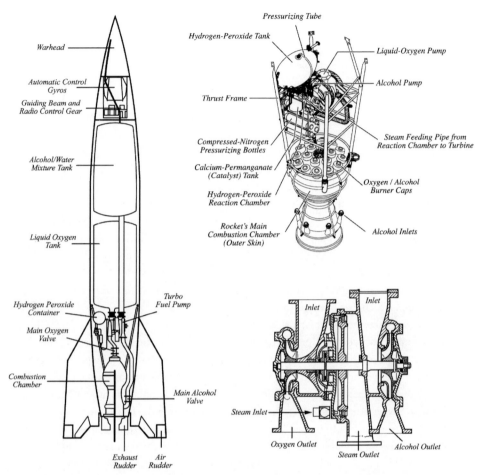

Diagrams of V-2 components including the engine assembly and turbine. The large propellant tanks were stacked inline and monopolized the majority of the fuselage. The alcohol pipe carried fuel to the turbine through the middle of the liquid oxygen tank. The fuel was channeled around the V-2's combustion chamber for cooling purposes. The first V-2 was fueled by a mixture of ethyl alcohol and water. The water reduced the rocket's overall thrust, but without it the engine temperature would have burned through the combustion chamber. Later, methyl alcohol was used to obtain greater range. The V-2's steering was controlled by gyroscopes which activated movable exhaust vanes for directional control.

powered by hydrogen peroxide, which was fed into the turbine by pressurization from compressed nitrogen. With the help of Klaus Riedel, Dr. Thiel refined the combustion chamber and nozzle design, which, combined with the turbo pump, gave the engine ten times more thrust.

Two large propellant tanks took up the majority of space inside the missile's main hull. Liquid oxygen was contained in the tank immediately above the engine and turbo pumps and was brought to the combustion chamber via a main valve assembly feeding 18 pipes down to the 18 injector orifices. The oxygen tank had a volume of 169 cubic feet and could contain over five tons of liquid oxygen. The fuel tank contained a mixture of ethyl alcohol and water was situated above the oxygen tank. The internal volume of this tank was 182 cubic feet, capable of accommodating over four tons of fuel, which gave the rocket approximately 60 seconds of powered thrust. The alcohol tank was tapered at the forward end to allow for the streamlined shape of the missile. The maximum width of the fuselage at the tanks was 64 inches. Alcohol was directed to the engine by means of a pipe that ran through the center of the oxygen tank. This diverged into two pipes and then again into three pipes at the bottom of the combustion chamber so that the fuel provided cooling by circulating the alcohol around the wall of the combustion chamber before it arrived at the injectors. The engine thrust frame was also an important component in the engine assembly. The frame was able to successfully transmit thrust to the rocket and withstand the punishment while remaining extremely lightweight. The turbine was driven by a chemical reaction between hydrogen peroxide and sodium permanganate. The hydrogen peroxide was forced into a decomposition chamber by compressed air where it was mixed with a small amount of sodium

(Top) The engine was supplied with propellants from high-capacity Walter turbo pumps. The pumps were driven by a turbine using a chemical reaction between hydrogen peroxide and sodium permanganate. The hydrogen peroxide was forced into a decomposition chamber by compressed air where it was mixed with a small amount of sodium permanganate solution. The resulting steam created from mixing the two chemicals generated high pressure, which rotated the turbine at 4,000 rpm and pumped the propellants to the combustion chamber at 33 gallons per second.

(Bottom) The V-2 engine assembly attached to the sturdy thrust frame. The frame was able to successfully transmit thrust to the rocket and withstand the punishment while remaining extremely lightweight. Using alcohol mixed with water for fuel and liquid oxygen as the oxidizer, the engine generated over 25 tons of thrust. The big combustion chamber featured an injector system at the top with 18 separate orifices through which alcohol would be brought to mix with the liquid oxygen. The engine was complicated, which made mass production more difficult. (Ed Straten)

permanganate solution. The steam resulting from mixing the two chemicals generated high pressure, which rotated the turbine at 4,000 rpm and pumped the propellants to the combustion chamber at 33 gallons per second. A heat exchanger ran from the turbine to one side of the tail of the rocket to vent the resulting exhaust gases near the fins.

The angle of the four carbon rudders in the engine exhaust blast would control the rocket. In conjunction with the gyroscopes in the control compartment and the four aerodynamic vanes located on the outside of each fin, the rudders provided primary control for the directional flight path. As archaic as it may seem now, with high-speed computers and satellites guiding today's modern ballistic missiles, the A-4 system of control surfaces backed by servos was a major breakthrough in rocket guidance.

Above the main fuselage and tank section of the A-4 was the forward control compartment. Tapering to the nose, the compartment was about 4.5 feet long and divided into four sections internally by four plywood sheets set at 90 degrees to each other. Each individual section housed vital flight-control equipment or electrical apparatus. Electrical power for the rocket was supplied by two 16-volt batteries, and one 50-volt battery provided the power to relay the gyroscopic signals to the servo mechanisms used for flight control. Other important items in the forward compartments included ground power control sockets, alternators, and three compressed-air bottles used to pressur-

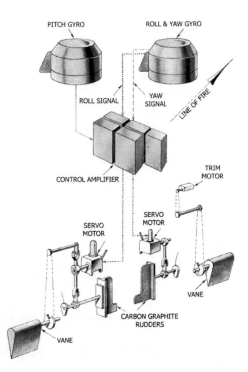

V-2 Control Surfaces and Mechanism. The steering of the rocket was controlled by two gyroscopes operated from a 500 cycle three-phase supply obtained from a motor alternator running from the lead-acid battery. One gyro was used to control the turning of the rocket into the trajectory and the other was used to control yaw and roll. The signals were channeled through the control amplifier and sent to the servo mechanisms mounted on a steel ring at the tail of the rocket. The figure above shows the outer vanes on the fins associated with the yaw and roll control and how they were directly linked to the associated carbon rudders through a system of mechanical levers and a chain drive.

ize the fuel tank before liftoff. Radio receiving equipment housed in one section interpreted signals from the ground for fuel cutoff at the appropriate moment. The two gyroscopes were responsible for azimuth and elevation angle, and when the missile deviated from the desired trajectory, the devices would bring each vector back on the predetermined line. Above the control compartment was the warhead. For one V-2, over 15,000 pounds of raw materials were needed (without the explosives or electronic devices). In the 1940s, the average price for a single rocket was approximately 119,600 Reichsmarks.

By June 18, 1940, all British forces had withdrawn from France. Both the Luftwaffe and the RAF had lost many aircraft and experienced pilots during the German campaign in the west. Over the next several weeks, the Luftwaffe began to replace their losses and moved their squadrons to the captured airfields near the coast. The delay also allowed the RAF to replenish their supply of available pilots and aircraft. British naval forces controlled the sea lanes, and in order for Hitler to invade Britain, the Royal Navy and the RAF would have to be subdued. The Luftwaffe needed to gain control of the skies over the English Channel before any planned invasion fleet could set sail across the Channel. On July 10, the Luftwaffe began attacking radar installations and shipping convoys near the south coast and Channel ports. Believing the British radar sites had been effectively disabled, the Germans began attacking RAF airfields on August 13, 1940. The objective was to keep constant pressure on the airfields and to destroy the RAF on the ground or in the air. The raids came night and day, and soon the RAF squadrons found themselves in a desperate situation. Many RAF airfields were destroyed, and the vital logistics of supply, maintenance, and communication were breaking down. However, in the face of this onslaught, the British pilots hung on and continued to fight gallantly.

Suffering from the lack of needed resources, progress at Peenemünde was slowing significantly. Languishing third on the priority list for available materials, many important projects were put on hold. Dornberger juggled whatever he could, borrowing men and material from construction allocations, just to keep development work on track for critical A-4 technologies. The materials originally designated for Peenemünde were being rerouted to projects with a higher priority status, such as U-boat and aircraft production. In October of 1940, Peenemünde officials complained to von Brauchitsch, who was now a Field Marshal, about their current situation, saying, "It is impossible to schedule or complete ongoing research with our priority status constantly fluctuating."

After an audacious attack by the RAF on Berlin, Hitler made the mistake of discontinuing the strikes on British radar stations and airfields and ordered his bombers to strike London in retaliation. Germany's daily bombing raids on London began in September of 1940. The Londoners were resilient in the face of the German bombers. Winston Churchill decreed the British "will never surrender" as the Battle of Britain unfolded in the skies above them. By taking the pressure off of the RAF, Hitler allowed the British to repair the airfields and regain the initiative in the air. Over a period of weeks, an increasing number of German aircraft and aircrews were shot down as the Luftwaffe slowly lost air superiority. Through his reports and discussions touting the usefulness of rocket research, and considering the general situation for Germany's armed forces at the time, Dornberger was finally able to capture the attention of the Nazi leadership. Dornberger pointed out that German bombers were being shot down over England after only five to six missions, and Dornberger argued this meant only six to eight tons of bombs actually landed on England before an aircraft was lost. Each bomber, along with the cost of training the crew, was 30 times more expensive than a single A-4. Thirty rockets could deliver more explosives for the same cost as one bomber, almost 25 tons more. Without any real evidence, Dornberger also suggested Germany may not be alone in its rocketry research and warned that the Americans, under Dr. Robert Goddard, might be only a few years behind in their research.[4]

In fact, Professor Goddard had made considerable progress, successfully launching several rockets to altitudes reaching 8,000 feet controlled by gyroscopes and moveable air vanes that corrected course in powered flight for almost 25 seconds. However, Dr. Goddard was working alone, without government sponsorship or involvement.

The daily night attacks on London, known as The Blitz, lasted until November 13. It was Hitler's first military subjugation and on November 17, 1940, he ordered the cancellation of the planned invasion of Great Britain. The Luftwaffe's inability to sustain a continued attack against Great Britain may have altered Hitler's attitude somewhat in regards to the rocket, considering that in March of 1941, he awarded Peenemünde a high-priority level status. However, this was for research and development of the rocket only. The planned series production facility received a lower priority. The rocket was still unproven, and the Nazi leadership wanted to find out if the rocket would "fly" before allocating the materials needed for full-scale production.[5]

Because of new proposals put forth by Munitions Minister Fritz Todt to restrict Peenemünde construction even further, Dornberger and von Braun, at the prompting of Field Marshal von Brauchitsch, flew to East Prussia to meet with the Führer at his headquarters on August 20, 1941. Also present at the meeting were General Fromm and Field Marshal Keitel. Dornberger made a presentation highlighting von Braun's work, development progress at the center, future projects, and the requirements necessary to complete their work on schedule. The objective of the meeting was to convince Hitler of the immediate need for priority funding of not only rocket development but also rocket production. This wasn't accomplished.[6]

After the meeting Hitler commented, "This development is of revolutionary importance for the conduct of warfare in the whole world. The deployment of a few thousand devices per year is therefore unwise. If it is deployed, hundreds of thousands of devices per year must be manufactured and fired."

Hitler still had no understanding of complex missile systems. His ridiculous pronouncement for suggested rocket stockpiles in six-digit figures was far in excess of the available resources. Nevertheless, the meeting did succeed in securing continued funding for the manufacturing plant at Peenemünde, without the interference of Minister Todt. The whole of Peenemünde was put in the category of "special enterprises," which provided further protection by combining all programs into one entity.

In late 1941, using the cover name of Versuchskommando Nord (Experimental Command North), Field Marshal von Brauchitsch authorized the withdrawal of thousands of technicians from the front lines and sent them to Peenemünde for continued research. In reality, the severe labor shortage in Peenemünde prompted Dornberger to lobby von Brauchitsch for the creation of the Versuchskommando Nord (VKN). In a show of commitment to Dornberger and the A-4 program, von Brauchitsch increased the overall availability of manpower in Peenemünde by ordering the formation of the new unit. This decision made by von Brauchitsch demonstrates his belief in the A-4. He thought it to be a decisive weapon, one that could possibly bring victory to the German people. If not for his unwavering support of Peenemünde and his determination to minimize the effects of what he saw as ridiculous bureaucratic contrivance, the progress of the program would have been considerably slower.[7]

By November, the VKN formation was complete. More than 600 officers and soldiers made up the new unit. Ostensively, the formation was a way to bring in more qualified personnel to help with the massive amount of work remaining in the development of the rocket and its systems. Formed on paper as a combat unit stationed on the homefront, the strange mix of chemists, electricians, and engineers was divided into six companies under the command of one sergeant major. Many of the men filling the ranks of this unit were from units in active combat. From out of nowhere they received orders to report to a place called Peenemünde. Most had never heard of it.[8]

The Peenemünde team forged ahead, even though in the back of their minds there was a real concern whether the A-4 would ever be accepted for mass production. Work had continued under the umbrella of German Army supposition for two full years, and by 1942, the first prototype missile was almost ready. All the while Dr. Todt, head of the Munitions Ministry, did his best to cut the flow of raw materials to the rocket facility. Todt was killed on February 8, 1942, when his aircraft crashed upon returning from a high-level meeting with Hitler at Rastenburg. The Führer took the news very hard, as Todt was one of his favorites. Even though progress on the A-4 had suffered under Todt's leadership, it was undeniable that armaments production had increased dramatically under Todt's supervision. Hitler wasted no time in naming his replacement. Albert Speer was appointed as Minister of Munitions the same day. Speer, only 36 years old at the time, was a rising star in Nazi circles. Hitler had complete confidence in the young man's abilities. Speer was regarded as complex, brilliant, and composed far beyond his years. In 1934 he was chosen by Hitler as architect for the construction of the Nuremburg Congress Stadium and then later the Reich's Chancellery in Berlin.

The buildup for the coming Russian campaign consumed the majority of Germany's available raw materials. The A-4 project was sidelined again as steel rationing was in effect and designated for other types of armaments production. Peenemünde encountered many difficulties in acquiring needed equipment such as heavy machine tools for rocket manufacturing. A few weeks after Speer's appointment, Dornberger redoubled his efforts to get rocket production underway. In the fall of 1940, Speer had been charged with the completion of the remaining Peenemünde construction projects; and on several occasions prior to his appointment as Minister of Munitions, Speer had met with Dornberger to discuss the lagging research at Peenemünde. From the time he first learned of Peenemünde, Speer had gained a fascination for the rocket center and its ambitious technicians with their fantastic visions. When he was young, he originally wanted to become a mathematician, though he ended up following in the footsteps of his father and grandfather and studied architecture instead.[9]

Dornberger recognized in Speer's appointment another opportunity to change the adverse opinions held by the Nazi leadership about the validity of rocket research. He moved quickly and detailed to Speer a plan in which large-scale production on the A-4 would begin by the fall of 1943. He co-opted the support of Colonel Zanssen, chief of the Peenemünde Army Experimental Station (Peenemünde), and together with von Braun, the three men lobbied the Ministry of Munitions with renewed fervor. Optimistically, Dornberger envisioned an attack on England as early as January of 1944, with missiles being assembled in Peenemünde and at facilities belonging to Hugo von Eckener's Zeppelin works at Friedrichshafen.[10]

On February 25, 1942, the first test prototype of the A-4 missile was complete. On a cold winter morning, Versuchsmuster 1 (Test Vehicle 1, or A-4/V1; not to be confused with the V-1 flying bomb) was moved into Prüfstand VII to begin initial testing. The purpose of this first prototype was for static testing only and by March 18, 1942, the rocket was ready to undergo its first firing test. Test Vehicle 1 featured a large steel corset that was manufactured to clamp the missile into the mobile test stand. Seen in many blueprints and sometimes mistakenly described as a V-2 variant with sinister intentions, the corset was merely a means of securing the device for static testing. The designers of the corset did not take into account the shrinkage that would occur when the cold liquid oxygen was pumped inside the missile's tank near the clamping device. The engineers watched in horror as the rocket assembly slipped out of the corset and crumpled in a heap at the base of the mobile test stand, thus destroying A-4/V1. Some have incorrectly hypothesized that this blueprint showing the A-4 "korsett" proves intent by German engineers to produce a V-2 variant which would carry atomic materials—a dirty bomb—to England. In fact, the aerodynamic configuration of this design was unstable.[11]

Operation Barbarossa, the invasion of the Soviet Union, opened on June 22, 1941. Field Marshal von Brauchitsch was forced to cosign and implement Hitler's infamous Commissar Order for guerrilla warfare, which essentially condoned the mass extermination of captured Soviet prisoners of war and civilians. By October of that year, the German armies were approaching Moscow. The German blitzkrieg was as devastating in Russia as it had been elsewhere in Europe. The campaign was truly a war of annihilation, and by the end of the first week of the invasion, 150,000 Soviet soldiers were either dead or wounded. Hitler was overconfident due to his rapid success in Western Europe and considered the Red Army to be easy prey. He reasoned a quick victory against the Red Army would encourage Britain to sue for peace. He expected victory within months and did not prepare for a war lasting into the winter. The turning point of Operation Barbarossa was when German armies advanced within sight of the spires of the Kremlin in late 1941. It was as close as they would ever get, for Stalin's troops defended Moscow ferociously and drove the Germans back into the frozen wasteland of Russia as the winter approached. Hitler blamed von Brauchitsch for the failure to seize Moscow, heaping so much abuse on the Army Chief that he suffered a heart attack. In failing health, von Brauchitsch was unceremoniously relieved of command and retired

Test Vehicle 1 (A-4/V1) seen just outside the south wall of the Test Stand VII arena during maneuvering trials in March of 1942. The camera crew on the left is filming the maneuvers, and the footage will be part of the compilation shown to Hitler on July 7, 1943. The rocket is resting on an early model of the Meillerwagen erector trailer. This version proved to be problematic and was later redesigned. The towing vehicle is the halftrack prime mover, the 12-ton Sd.Kfz.8, with a full complement of soldiers. Test Vehicle 1 was destroyed a few days later during a mishap inside of Test Stand VII. (*Deutsches Museum Munich*)

in December 6, 1941. The next day, Hitler took over as Army Commander in Chief. He now had absolute power over the fate of Germany's armed forces. Like an omen of future events, that same day, Japanese naval forces attacked Pearl Harbor, thus bringing the United States into the Second World War.

A steady drone of bombers was heard approaching the German sea port of Lübeck on the evening of March 28, 1942. The RAF was about to embark on a new aerial bombardment strategy in which entire German cities would be targeted for destruction. By the following morning, hundreds of German citizens were dead, and the city was laid waste. Hitler was enraged and demanded retaliation. But with the Luftwaffe's defeat over England, there was little recourse for revenge. Hitler met with Speer, who happened to have a copy of Dornberger's new memorandum that proposed a hardened rocket-launching bunker to be constructed on the coast of France. Speer had previously asked Dornberger to outline his own aspirations for the rocket program, and in the memorandum, "Proposals for Employment of the Long-Range Rocket," Dornberger touted the benefits of the rocket and discussed the ineffectiveness of other conventional weapons. In the memo, Dornberger reasoned it was feasible that almost 5,000 missiles per year could be launched against Allied targets such as British industrial centers, seaports, and the city of London itself. For the first time, he gave an indication of what the psychological effect might be on the civilian population during this missile bombardment. Touting the advantages of using missiles, he pointed out the targets could be attacked night or day, regardless of the weather conditions. The memo was highlighted in Hitler's mind a few weeks later when a thousand RAF bombers attacked the city of Cologne. Even though he demanded retaliation, Hitler still refused to give the rocket production priority.

Hitler's lust for revenge against England was not tempered by any understanding of reality. Unbelievably, after starving the rocket program for years, Hitler suddenly demanded that when possible, a rocket offensive against England should commence, and he wanted no less than 5,000 rockets to be launched at the outset. Dornberger was flabbergasted. Maybe it was his own fault for mentioning 5,000 rockets in his feasibility memorandum, but there was no way current manufacturing, without priority status, could meet the quota of monthly rockets demanded by Hitler; and even if the rockets were available, there were not enough liquid oxygen manufacturing facilities to meet the requirements for such an initial onslaught. There were no trained troops to deploy the weapon and no support equipment to outfit such battalions. The rocket was yet to even fly.[12]

Despite Hitler's vacillations, time dictated that plans for operational deployment of the missile should commence. In April of 1942, a command structure was set up within the Versuchskommando Nord (VKN) charged with the responsibility of developing, testing, and evaluating field-support equipment along with creating instructional text and tables of organization for the rocket units. A multistage training program for potential rocket personnel was created. The introductory portion of the training lasted eight days, and it gave each potential recruit an overview of the missile. Afterward, the trainee was given two parallel seven-week courses. One portion was dedicated to basic electronics, and the other illustrated various technical aspects of the missile. After their instruction to this point, the trainees were given oral and written exams. If they passed, they moved on to Peenemünde for further training. However, if they failed, they were washed out.[13]

All the while, work on Test Vehicle 2 (A-4/V2) progressed, and the rocket was ready on April 29, 1942. The following week an extensive series of fuel-injection tests were run on the new rocket. On May 14, 1942, the engine was successfully fired with the rocket mated in one of the giant mobile test stands in Prüfstand VII. The following morning, an RAF reconnaissance aircraft was seen circling over the Usedom peninsula. Test Vehicle 2 was still in the arena as the high-flying British Spitfire made several lazy passes over the rocket center. The German engineers could only wonder if the pilot was suspicious of the goings on below. Testing for A-4/V2 encountered a setback on May 20 when an accident damaged the rocket. The damage was serious enough that Test Vehicle 2 had to be taken down and disassembled to make repairs. The rocket returned to Prüfstand VII on June 8 and was readied for firing on June 13.

Early that Saturday morning, hundreds of engineers, technicians, draftsmen, and military personnel were watching from rooftops and in the streets, all in anticipation of the first flight of the big missile. The rocket ignition occurred at 11:52 AM, and for the first time, the A-4 rose from Peenemünde. It started out vertically, but just as some on the ground began their cheers, it began oscillating—the black-and-white checkered pattern painted on the rocket clearly illustrated the missile was rolling back and forth, to the left and then to the right. This wasn't planned. Soon the wobbling object was rotating faster and faster; the tilt program came to bear, trying to maneuver the rocket into a ballistic attitude, and this made things even worse. The fuddled missile disappeared into the clouds with its engine still roaring. Although no one could see it, the rocket passed through the sound barrier and continued on until the fuel was exhausted. Then the silence was deafening. All eyes searched the skies. Almost a minute later, the huge missile tumbled out of the cloud cover and finally fell into the Baltic, sending up a tremendous splash of water.[14]

The next A-4 test flight was even worse. On August 16, 1942, Test Vehicle 3 (A-4/V3) was readied for launching in Prüfstand VII. On this day some very prominent officials would be on hand in Peenemünde. The armaments chiefs of three service branches, including Field Marshal Erhard Milch, who had just ordered top priority for the Luftwaffe's development of the flying bomb, were to be in attendance. Also traveling to Peenemünde to witness the test was Minister of Munitions Albert Speer, although even as the countdown commenced, Speer had not yet arrived. The rocket lifted off just past noon but experienced an immediate power failure that shut off the transmitter. However, the rocket stayed on course and climbed steadily, passing through the sound barrier. Speer watched the missile's flight from his aerial vantage point as his aircraft came in to land on the airfield at Peenemünde West.[15]

Internal mechanical failures and problems with the electrical system caused the A-4's engine to shut down prematurely at a height of about seven miles, and the deceleration caused the remaining fuel to slam into the top of the propellant tanks, which broke the missile apart. The remnants fell into the Baltic waters approximately 5.4 miles downrange.

The thunderous spectacles were certainly breathtaking to all who witnessed these test flights, but when would the missile demonstrate its true capabilities? To some observers, the A-4 seemed to be a very exciting, but extremely expensive, giant firecracker. In fact, Dornberger wondered if the detractors might even be correct. Would this damn thing work? After Germany had spent millions of marks on the missile program, all during a time of war, he had little choice but to start another intensive search for the possible cause of the failure. These failures were to be expected when bringing the burgeoning technology of a sophisticated rocket to the battlefield, but for the sake of survival, they had to get it right—and soon.

August 16, 1942, Test Vehicle 3 (A-4/V3) is undergoing prelaunch preparations in the middle of Test Stand VII at Peenemünde. This was the first A-4 rocket to exceed the speed of sound; however, an internal mechanical failure caused the rocket's engine to shut down prematurely and the deceleration caused the remaining fuel to slam into the top of the propellant tanks, which broke the missile apart. (*NASA*)

October 3, 1942, was a cool but sunny day in Peenemünde. Launch preparations for A-4/V4 had, so far, gone without a hitch. As the missile stood upright in the center of P-7, frost was forming on the outer skin as it was filled with liquid oxygen. A successful launch was anticipated by everyone; however, there remained some doubt and a little apprehension. During three previous attempts, the missile had failed. Despite the hard work of the rocket team, rumors were quickly spreading that if this last test met with failure, the Army's rocket program would be dismantled and all Peenemünde personnel would be sent to the battlefront! Clearly, this was an important day for Peenemünde's survival.

Technicians made the final checks on the 47-foot-tall rocket, and then the immediate launch area was cleared. Dr. Thiel, as head of the launching complex, was inside the control bunker along with the launching personnel. Away from the arena in the Ost Entwicklungswerk (East Development Works), Dornberger, standing on the roof of the Measurement House with Dr. Zanssen, gave the order to commence the countdown. In the distance they could see Prüfstand VII surround by a forest of conifer trees. Farther on was the little island Greifswalder Oie. Dornberger would have a clear view of the missile's flight path from this vantage point. To monitor the ignition, there was a television installed on the roof along with a Doppler receiver, which would give Dornberger a clear indication of the missile's speed throughout its trajectory. Dornberger could see von Braun, Dr. Steinhoff, Commander Stegmaier, and other familiar faces on the roof of the works assembly building not far away. Also, he recognized Dr. Hermann and Dr. Kurzweg, who came over from the supersonic wind tunnel to watch the launch. People exited their workshops and offices; hundreds were standing in the streets, all excited to witness the launch. Loudspeakers positioned around the complex announced the countdown. Observers were waiting in nervous anticipation as the critical portion of the flight timetable was fast approaching.

Like a monolithic statue to the rocket team's hopes, the rocket glinted in the afternoon sun, still connected by an umbilical cord to the instruments in the control bunker. Full of fuel, the rocket now weighed more than 28,000 pounds. As oxygen vapor vented from the rocket, the final check of the electrical system, valves and lines was completed.

"X minus three minutes," came from the loudspeakers.

The vapor stopped when an operator in the control bunker closed the venting valve. Pressurization of the oxygen tank began. The alcohol tank was already pressurized; the gyros were spinning in the control compartment; everything was now ready for ignition. Dornberger looked intently at the television screen.

"X minus one."

Ten seconds before ignition, a green smoke flare was fired nearby to profile the wind conditions. All heads looked up for a second, and then knowing glances were exchanged by those standing close to one another.

"3 . . . 2 . . . 1 . . . Zundung" (ignition).

The high-speed recorders and the oscillographs were running as the engineers hit the preliminary ignition switch. Sparks appeared under the rocket as a bright orange flame began leaping out from the rocket chamber. The turbo pump was idle at this point during Vorstufe (preliminary stage). Flowing under gravity, the fuel only generated marginal thrust (16,000 pounds), not enough to lift the fully fueled 15-ton rocket. When the launch control officer in the bunker saw that the flame was burning evenly, he flipped another switch and announced "Hauptstufe" (main stage). After three seconds, the umbilical cable fell off and the turbo pump was turned on. Rotating at 4,000 rpm and generating 540 horsepower, the turbo pump screamed as it forced fuel into the rocket motor. A thrust of 50,000 pounds accelerated the rocket steadily skyward. The men in the control bunker and on nearby rooftops could feel the vibrations throughout their bodies as well as under their feet. The noise they experienced during previous static engine tests was tremendous, but those tests were directed toward a water-suppression pit. The cacophony created by an actual launch was startling to most, almost unbearable to some.[16]

Looking through his binoculars, Dornberger watched the rocket ascend. The jet of the exhaust was perfectly shaped, and he felt the critical moment had passed when he observed no rotation along the missile's longitudinal axis. The black-and-white roll-pattern painted on the rocket remained in position. At a height of one mile, the exhaust gases reached a velocity of 6,560 feet per second at over 5,000 degrees Fahrenheit, with the engine generating a flame about as long as the rocket itself.

The rocket slowly began to pitch over. Dornberger watched breathlessly as the accelerating rocket pitched 50 degrees, which would give it the ability to reach its optimum range. From the loudspeaker came a steady narration of the flight time in seconds—"14, 15, 16"—and a growing monotone hum coming from the acoustic measurement of the rocket's velocity. As the rocket climbed higher, so did the pitch of the signal. A remote microwave station located 12 kilometers from the launch pad transmitted the exact position of the missile.

How impressive this flight was! Continuing with a steady roar, the rocket gained more and more speed. Within a few seconds the rocket would break the sound barrier.

"Schallgeschwindigkeit!" (Speed of sound!) was announced.

Everyone hesitated in their excitement. This was a critical point. Could it be done? Some said it was impossible; little was known about the dangers of exceeding the speed of sound to a mechanical device. Less than ten seconds later, the rocket was still in good shape, having attained an altitude of 33,000 feet, and was fast approaching Mach 2—twice the speed of sound. A few seconds later, a white vapor trail formed, which led some people to believe an explosion had occurred. However, the steady drone of the velocity meter assured Dornberger that the rocket was intact. The sudden appearance of the contrail against the background of the blue sky was something the scientists had not experienced. Some feared an explosion or malfunction had occurred, however, it was only a vapor trail formed at high altitudes as the exhaust gases condensed due to the low pressure and the low temperatures. Later, it would become a common point of measurement, "the frozen blitz" as the Germans later called it. (All this was before contrails became a deadly augur of the American daylight bomber offensive over Germany.)

At around 50 seconds, engine cutoff was approaching. The preprogrammed gyros were busy making adjustments. The autopilot system would shut off the engine when it sensed the correct azimuth, velocity, and tangential attitude for the trajectory. A transponder in the rocket sent the position and attitude down to a ground-based radio station.

A few seconds later, observers on the ground called out, "Brennschluss!" (Engine shut down!).

At a distance of about 18 miles downrange, Dornberger could still see the missile, and he knew it was continuing to gain speed by the pitch of the velocity meter. Looking through his binoculars, he saw a thin stream of smoke exiting the nozzle of

Liftoff of the first successful A-4 on October 3, 1942. This rocket, Test Vehicle 4, approached 3,000 mph and splashed into the Baltic Sea after rising to a height of over 50 miles. The timer inset on the right records the duration of the flight on film for post flight analysis.

the engine. The flow of steam into the turbine was decreased when the main hydrogen peroxide valve was closed, but the eight-ton valve remained open. This dramatically decreased the level of thrust from the engine and permitted a more exact moment of thrust termination. It was easily seen from the ground as the steam of exhaust became smaller. Now the A-4 was traveling at more than 3,000 miles per hour with the red-hot jet vanes glowing in the heavens.

As the velocity meter continued its drone, the distant roar of the engine faded. Dornberger took a moment for a deep breath, and then a feeling of relief washed over him. Success! After ten long years, they finally had succeeded. Dornberger was overcome. Lost in the reality of what he had just witnessed, he was speechless. They had just propelled an object into space. Emotions of joy washed over him. Feeling much the same, Zanssen turned and extended his hand in congratulations. Dornberger took hold of his hand and shook it, then without letting go, he pulled Zanssen close and embraced him. The two men hugged and yelled in jubilation. As they looked around, it was evident that everyone else was also celebrating.

Dornberger quickly came down from the roof. Exiting the front of the building, he quickly shook hands with a few colleagues and then grabbed von Braun by the arm. Pulling him along, they jumped into Dornberger's car and sped off toward P-7 to examine the launch site and confer with the launching personnel. All the while, the rocket continued its journey over the Baltic, the loud drone of the velocity meter could be heard echoing throughout the complex.

As the entered the arena, they could see everyone had gathered around Dr. Thiel. The boisterous group was shouting, laughing, and celebrating. Each man wished to inform Thiel about what they observed during the flight. Dr. Thiel, with his ever-pres-

Unexpectedly, on December 11, 1942, Reichsführer SS Himmler visited Peenemünde. Himmler (wearing glasses in center) listens to Colonel Dornberger (fourth from right) explain a static engine test. General Leeb (second from right) keeps an eye on Himmler. Himmler was interested in spreading his influence within the burgeoning rocketry program. (*Deutsches Museum Munich*)

ent unlit pipe in his mouth, was not the kind of man who would normally show his exuberance. However, when he and Dornberger shook hands, the excitement was written all over Thiel's face. They immediately began discussing numerous ways to improve on their success. There was great emotion all around. Dornberger hushed the group and instructed them to listen. The test was not over. The rocket would reenter the earth's atmosphere at any second. They listened to the pitch of the velocity meter. At around 300 seconds, the sound disappeared. The rocket was down. A-4/V4 soared perfectly into the heavens and splashed into the Baltic 190 kilometers downrange, only a few miles from its intended target area.

That evening Dornberger held a small party at his residence. The celebration was raucous with plenty of beer and wine all around. At an opportune moment, Dornberger stood and gave an impromptu speech.

"History of technology will show that for the first time, we have invaded space with our rocket. Your names, my friends, are associated with this feat. We achieved this feat with an automatic control system. We have demonstrated that an unmanned or manned air vehicle can break the sound barrier. Mark this well, we have used space as a bridge between two points on the earth; we have proved rocket propulsion practicable for space travel. This third day of October, 1942, is the first day of space travel. However, as long as there is war, our efforts will be used to produce a weapon. I hope this achievement will allow us to continue our work."[17]

The hard work had paid off. After the successful flight, von Braun was awarded the Kriegsverdienstkreuz I Klasse mit Schwertern (War Service Cross, First Class, with Swords).[18]

This was only one rocket however, and as the A-4 test program continued into 1943, the success would prove difficult to duplicate. Technicians would encounter numerous problems in subsequent weeks. The next dozen test shots ended in failure, some in spectacular fashion, with fiery crashes near the launch site. On January 7, 1943, test missile A-4/V10 had risen from its launch table only a few inches when an explosion in the combustion chamber caused the missile to topple. The fuel tanks burst open as the middle section slammed into the floor of the arena, and the ensuing explosion sent a huge fireball into the air over Prüfstand VII.

The center of Test Stand VII as seen today. The marker shows the spot where so many test rockets were launched, including the first successful shot on October 3, 1942. Immediately after the war, the Red Army, in accordance with the Potsdam Agreement, demolished all installations at Peenemünde. The large German experimental center was razed to the ground. After 45 years behind the Iron Curtain, the forest growth and abundant wildlife has reclaimed the area. (Author)

Ever increasingly, news of the A-4 progress reached the higher echelons of Nazi leadership. Unexpectedly, on December 10, 1942, Reichsführer SS Himmler telephoned to say he would be visiting Peenemünde. The news of Himmler's visit bothered the Army Weapons Office. Colonel General Fromm and General Leeb hurried to meet with Dornberger in Peenemünde. When Himmler arrived the next day, he was accompanied by SS General Emil Mazuw, who was the police commissioner from Stettin and also the local SS representative for the Baltic. Dornberger gave an informative lecture about the rocket and a few static rocket engine tests were demonstrated. Himmler listened and watched attentively, asking questions from time to time.

A-4/V9 took off from Test Stand VII the next morning. Four seconds after liftoff, a massive explosion ripped the tail section off the rocket, and it crashed nearby. After the excitement, the group set off for lunch in the mess. Relaxing after his meal, Himmler explained the purpose of his visit. "With the recent progress of your rockets, there has been much talk about it in the Führer's inner circle. Soon your work will become the concern of the German people and no longer the affairs of the Army Weapons Office or the Army itself. The Führer is about to give your project his support, and that is why I came."[19]

Dornberger, although intrigued by the statement, knew Himmler had been closely watching rocket development. Himmler had even established an experimental rocket testing station of his own, the SS facility at Grossendorf, also on the Baltic near Danzig. Specializing mainly in the development of solid-fuel rocket artillery, it was under the direction of former Raketenflugplatz member SS Captain Rolf Engel, who held a long-standing grudge against Army Ordnance. Engel, who previously had also worked at Kummersdorf, disliked Dornberger for unknown reasons.

V-2

"I want to protect your work against sabotage and betrayal within your ranks," said Himmler.

Himmler's statements, insulting to the Army officials present, were seen for what they were. General Fromm, sitting next to Himmler, shot back, "Reichsführer, we certainly would welcome more security in the outlying areas around Peenemünde, but since the research center is an Army establishment, it is the Army's responsibility to provide security from within."

Colonel Dornberger was surprised at General Fromm's statement but also relieved. No one spoke to Himmler that way. Himmler was silent for a few seconds, reassessing his surroundings and concluding there was no need to press things now. Himmler feigned agreement. "General Mazuw, you will take responsibility for setting up a restricted zone around Peenemünde."

That evening Dornberger escorted Himmler to his aircraft at Peenemünde West. As Himmler took his leave he said, "I am excited about the success of your work, it interests me greatly. I offer you my help. Very soon I will come again, the next time by myself so we can have a good conversation together and also with your colleagues. I will phone you to let you know in advance of my next visit."

The success of the Army's A-4 program and Dornberger's suggestion that the A-4 could compensate for the Luftwaffe's inability to sustain a bombing campaign against Britain provoked the Luftwaffe to begin its own research program into missiles in 1942; namely, the Fieseler Fi-103 cruise missile, also known as the FZG-76 and later known as the V-1, or flying bomb. The Fi-103 was a very simple, almost crude, weapon compared to the sophisticated A-4. Propelled by an engine called a pulse-jet, which received air in at the front and expelled combustion gases from a long exhaust tube, the flying bomb would cruise in a low trajectory to its target, whereupon the engine would shut off, and it would fall to the ground near its objective. The engine featured a series of louvers in the front that would open to let air rush in and then close when the appropriate pressure in the combustion chamber had built to a sufficiently high level for fuel to be injected. An igniter would trigger an explosion, or pulse, which would send the hot gases through the rear exhaust tube. The engine pulsed at a rate of over 500 explosions per minute, giving the weapon a very distinctive sound when moving through the sky. At its introduction, the Army worried that the Luftwaffe might try to kill the A-4 project by substituting the flying bomb in place of the A-4 altogether.

In late 1942, the new German Armaments Minister, Albert Speer, convened a commission of senior officials to look carefully at the merits of both weapons. In May of 1943, the commission recommended that both programs proceed. The flying bomb was not particularly accurate, and its speed, although faster than many fighter aircraft, was still modest in comparison to the A-4. Flying in a straight line, it used an onboard meter to measure the distance traveled before the engine shut off and it fell to earth. It was vulnerable to antiaircraft fire, barrage balloons, and the newest British fighter interceptors. However, it was a remarkably cheap and efficient means of delivering a bomb on target, and with a range of nearly 190 miles, it could compete with existing aircraft for many of the important objectives. Moreover, with a ton of explosives in the nose, it would act as an unmanned bomber, thus reducing the high attrition rate experienced by the Luftwaffe flight crews. On the other hand, the A-4 would be impervious; there was no countermeasure for it. Once launched, the missile would make a five-minute journey into near space and impact its target at a speed of more than 1,500 miles per hour with no warning.

On December 22, 1942, Speer was summoned to attend an important meeting at the Ministry of War in Berlin. During the course of the meeting, the subject of the A-4 was raised by Speer. He was able to convince Hitler to finally sign the order to expedite mass production of the rocket. Speer also reminded the Führer of Dornberger's proposal of a hardened Blockhaus (block house, concrete bunker) in northern France from which the rockets could attack England. Later in the day, Speer ordered the beginning of normal manufacturing procedures for the A-4. This practice involved setting up development commissions composed of industry experts who were to analyze manufacturing techniques and recommend improvements.[20]

Speer began preparations for the launching bunker in late December by contacting Peenemünde and requesting an immediate survey of the Channel coast to find an appropriate construction site. Dornberger was away on Christmas holiday and learned of Speer's request later. In the months of December and January, officers, geologists, engineers, and scientists under the direction Major Thom from Peenemünde, Dornberger's chief of staff, scoured the countryside of northern France searching for an appropriate site to construct the massive launching bunker.

On January 8, 1943, Dornberger traveled again to Berlin to see Speer. He was informed by Speer that Organization Todt would be working closely with Peenemünde during the construction of the launching bunker, which was to be built in the Pas de Calais of northern France. He also introduced Colonel Dornberger to a man named Gerhard Degenkolb, a man who would help organize the mass production of the rockets. Degenkolb was the former director of the Demag Engineering Works and the man responsible for streamlining the production of Germany's locomotive industry. Speer touted Degenkolb as a miracle worker when it came to managing important large industrial projects with seemingly impossible deadlines. Dornberger's first impression of Degenkolb was not positive. The balding Degenkolb seemed to be boastful and pretentious. Without any previous knowledge of the rocket program, he proceeded to tell Dornberger how he intended to move the rocket production forward. Degenkolb was already talking about setting up a production committee, and this made Dornberger very nervous. In fact, Speer had already named Degenkolb as head of the Special A-4 Committee. Nevertheless, Dornberger invited Degenkolb to come to Peenemünde to familiarize himself with the rocket and the facilities.

The following day the two men traveled north to Peenemünde. Degenkolb observed the progress on the A-4 in an effort to determine the requirements for a projected production schedule. The engineers had begun to outfit many of the test rockets with equipment that would help improve the efficiency in production and operational use. However, Degenkolb was unable to appreciate that the A-4 was still in development. He believed production plans should move forward immediately. Dr. Thiel spoke to Degenkolb about continuing problems with the big rocket motor, while von Braun pointed out that requirements for production would demand design modifications to the missile and stressed the need to continue with experimental tests to refine the device. The engineers at Peenemünde imagined larger, more powerful rockets in the future. The A-4 was seen as the forerunner to subsequent designs, each one building on the lessons of the other. Production of the A-4 would be a tremendous undertaking; however, Degenkolb was secure in his belief that full-scale production should commence immediately. After the tour, he returned to the Ministry of Munitions in Berlin completely confident that the A-4 could be transitioned into mass production quickly and efficiently. One week later, Degenkolb was named chairman of the Sonderausschuss (A-4 Special Committee).

In previous months Dornberger and von Braun had been working closely with an engineer named Detmar Stahlknecht, a commissioner at the Ministry of Munitions, who had a great deal of experience with aircraft production. But without the priority status needed for production, almost nothing could be accomplished. Dornberger had sought higher priority, and now the Führer had given his support to rocket production and deployment, but was this what he wanted? Degenkolb? Only a few years before, Degenkolb—a dedicated Nazi—had publicly blasted the efficiency of the Army Weapons Department. He was instrumental in the reorganization of the armaments industry—a reorganization in which the influence of the armed services was sapped. He was a man with a ruthless reputation; he got the job done at all costs, no matter how much unrest might be the result. Dornberger was unsure. Maybe Degenkolb was the perfect man for the job of pushing forward A-4 production. Then again, maybe his appointment would lead to ruination.

The Special A-4 Committee conducted its initial meeting in February of 1943. With Degenkolb as director, the committee was organized into 21 subcommittees to cover the panorama of production for the more than 17,000 components making up the rocket. The committee's number-one goal was the completion of a single design, followed by the creation of production sites. It would also address the procurement of a large enough labor force to begin production. On top of this organization Hitler had created the Entwicklungskommission für Fernschiessen (Development Commission for Long-Range Rocket Bombardment). This morass of bureaucratic commissions, committees, and subcommittees, all influencing Peenemünde, was an organizational nightmare.

Dornberger's reservations about Degenkolb's appointment were justified. On February 3, the financial department of the Ministry of Munitions, at Degenkolb's urging, called Dornberger to Berlin. He was informed by the chief of the financial department that Peenemünde, because it was now going to be a manufacturing facility, should be transformed into a private company, whose stock would be owned by the state. Dornberger argued vehemently that this was unacceptable, they had no authority to "confiscate" the Army establishment, and that such a change would be disastrous for the rocket program. Finally, after some heated discussion, he was able to convince the financial chief to delay the decision.

The financial and engineering representatives of the Ministry of Munitions came to Peenemünde a few weeks later with an initial A-4 production schedule. Stahlknecht presented his own idea about the precise level of unit output. The pilot production facility at Peenemünde would be put into operation first, with a goal of 300 rockets each month beginning in January 1944; and then later, production from the Zeppelin Works at Friedrichshafen and the Henschel operated Rax-Werke at Wiener Neustadt in Austria would be added; and by September of 1944, the combined total output should equal 600 rockets per month. It seemed feasible to Dornberger, but if the A-4 was going to be in operation by the end of the year, even in limited quantities, there was not a moment to lose. Back in Berlin, Degenkolb initiated his cumbersome administration. He suspected that Dornberger was not putting sufficient emphasis on the imminent production of the rocket. Degenkolb sent Peenemünde his plan for a production start in October 1943 with 300 missiles per month rising to 900 per month in December. In his arrogance, he was confident that he could accelerate the seemingly reluctant technocrats into production status, but a schedule such as this was an impossible task. The technical difficulties that remained before production could commence were beyond Degenkolb's comprehension. Several weeks later, Speer was able to negotiate a compromise in which the Stahlknecht production proposal was reinstated.

When Dornberger had first proposed the missile production plant at Peenemünde, he envisioned it as being all-inclusive, with skilled laborers brought over from the Development Works. Even then, Dornberger had no idea where he would find enough skilled workers to support both facilities. The production plant would now be a "final assembly" location, and fewer skilled workers would be required if most of the main components were manufactured elsewhere. Almost 1,000 Polish workers, who were employed by Speer's Construction Group Schlempp, had been completing the various Peenemünde construction projects since 1940. By early spring of 1943, more than 3,000 foreigners were working in some capacity at Peenemünde. At that time the general consensus was for the use Russian POWs for the Peenemünde plant. It was a practice that was becoming widespread in German industry.

In April of 1942, Fritz Sauckel was named as Plenipotentiary for Labor Allocation. His task was to supply the manpower required for the armaments and munitions production program. Available German workers were almost nonexistent at the time, and the influx of POWs resulting from Germany's swift military victories presented a qualified labor source to draw from. As a result, millions of workers were seized in the occupied territories to work in German industry. Sauckel, who was later executed in Nuremburg on October 16, 1946, directed that workers were to be exploited "to the highest degree possible at the lowest conceivable degree of expenditure." The goal of the Nazi slave labor program was to strengthen the war machine while carrying out the Nazi objective of destroying persons they deemed inferior. Millions of forced laborers were deported to Germany and placed under conditions so inhuman that countless numbers perished. However, the decisions that led to the eventual use of concentration camp workers for rocket production were primarily made by the Special A-4 Committee, which approved the use of slave labor upon the recommendations of the labor supply subcommittee.

Reichsführer SS Himmler had already established a system of SS-run concentration camps for the extermination of Jews and subversives in the occupied territories. With the appointment of Sauckel, Himmler was persuaded to guarantee his share of the pie. He created the SS Economic and Administrative Main Office, in direct competition with Sauckel, and began leasing concentration camp laborers to German industry for a nominal fee. In April 1943, Arthur Rudolph, the Chief Production Engineer at Peenemünde, toured the Heinkel aircraft plant north of Berlin, which already used SS-supplied concentration camp laborers, and returned touting the benefits of possibility of using concentration camp laborers at Peenemünde.[21]

Industry officials found it appealing, as the SS would take responsibility for policing, feeding, and housing the prisoners. The following month, Sauckel was treated to an A-4 demonstration at Peenemünde and came away promising all of his support. But it was the SS Economic and Administrative Main Office that would deliver the first foreign laborers to several rocket component manufacturing facilities. On June 17, the first 200 prisoners, some who had been technicians or machinists before the war, arrived at Peenemünde from the Buchenwald concentration camp. A special workers camp called Trassenheide had been constructed just outside the rocket center, and SS guards maintained control over the prisoners. The detainees' first job was the completion of the security fencing around the F1 Building. A few days later, another 500 prisoners were delivered to the Rax-Werke in Austria. Throughout the summer, concentration camp laborers continued to arrive at the various production plants, including Wiener Neustadt and Friedrichshafen.[22]

In March of 1943, Organization Todt engineers descended upon the Foret d'Eperlecques (Eperlecques Forest) about five kilometers from the town of Watten (Calais, Saint Omer), France, to begin construction of the first enormous rocket-launching bunker. Plans for the bunker had been drawn up in January and February by experts from both Peenemünde and Organization Todt. This enormous concrete bunker was to accommodate V-2 reception and storage, preparation of rockets for launch in a sheltered and controlled manner, on-site production of liquid oxygen, and launch control with two firing pads. The project was code-named Kraftwerk Nord West (KNW).[23]

Eperlecques was chosen for several reasons. For the bunker to be logistically accessible during construction, and then later during operations, it had to be built close to a navigable canal for large barges and a railway linking Calais to Basle. The site would be sheltered somewhat by the location being at the foot of a forest that rose some 90 meters, just inland enough to be out of range of naval guns. The roads were excellent near Watten, and electrical supply was available as high-voltage power lines were already in place. There was no local railroad to the bunker area originally. The construction of a new 12-kilometer-long connecting railway was ordered on March 29, 1943. Nearly 6,000 laborers, Building Battalion 434, along with earth-moving equipment, moved into the site. The concrete rocket bunker would measure over 300 feet in length and 250 feet in width and rise to a towering height of 93 feet. A project of this size required an enormous labor force. Thousands of workers, mostly Frenchmen liable for the *service du travail obligatoire* (STO) (obligatory work service), but also Belgium, Dutch, and later, Russian prisoners, were forced to work under grueling conditions when constructing the bunker. The rate of progress was amazing due to advanced mechanized excavators, concrete plants and pumps, and the availability of forced labor. Sophisticated concrete pumps allowed a continuous flow concrete to be sent directly from the mixers toward the various points of construction. The building site was linked to where the materials were unloaded by a Decauville narrow-gauge two-way railway track. Near the town of Watten, workers unloaded the gravel, sand, and cement from the barges onto small trams that were going up and down the forest of Eperlecques constantly. The Decauville line traveled over a wooden bridge, through the forest, to the place where the concrete mixers were working. The concrete mixers received their supplies through the force of gravity, which helped speed the process. Using giant floodlights, laborers worked night and day during 12-hour shifts.[24]

On April 13, 1943, Oberstleutnant Thom, accompanied by representatives from Organization Todt, came to visit the construction site and took a series of panoramic photos of the works. The photos featuring the massive site with all of the concrete mixers, the Decauville tracks with little steam engines, the giant lifting cranes, and cavernous earthworks were shown to Hitler that summer. The deadline for completion of the project was set for October 31, 1943.

Dornberger was partially responsible for the interest shown by the SS toward the rocket program. In his attempts to push production forward, he co-opted the support of anyone who might be able to influence Nazi leadership. Several days after Himmler's first visit to Peenemünde, Dornberger asked Gerhard Stegmaier to send a letter to Himmler, which was conveyed through the SS Main Office, requesting Himmler to arrange an official A-4 presentation for the Führer. On January 23, 1943, Himmler approached Hitler and asked for complete oversight of the rocket program. Hitler refused Himmler's request for control and also denied the Dornberger's request for the A-4 presentation; however, Himmler's great interest in the rocket reinforced, in Hitler's mind, the potential decisiveness of the weapon. If people were noticing Peenemünde, it was because Hitler was increasingly speaking about it. If the Führer was warming up to the idea of the rocket program, Himmler certainly wanted the power of controlling it.

The next time Himmler paid a visit to Peenemünde, he arrived less conspicuously. He came all by himself, driving his own car, on the evening of June 28, 1943. Himmler was accompanied to supper, and afterward, a small group, including Himmler, Dornberger, von Braun, Stegmaier, Steinhoff, and a few others, retired to the sitting room. The chat began with the rocket men recounting their early days at

Kummersdorf. Himmler took the opportunity to ask more questions about the activities at Peenemünde. Eventually, the topic of conversation turned to politics, with Himmler reiterating the Führer's vision for Western civilization—one in which Europe was dominated by Germany. The following morning, after only a few hours of sleep, von Braun escorted the Reichsführer to the viewing area near Test Stand VII as A-4/V38 was being prepared for launch. Just past 9:00 AM, the rocket lifted off with a pronounced wobble. After climbing away from the immediate launch area, the engine stopped, the rocket turned over, and it began falling to earth in a ballistic attitude. Only 30 seconds after liftoff, it crashed less than two miles away, right in the middle of the adjacent Luftwaffe airfield, destroying three parked aircraft and blowing out the windows of nearby hangers. While inspecting the huge crater, which was almost 100 feet in diameter, Himmler remarked sardonically, "You seem to have a very effective close-range weapon here." That afternoon another A-4 was launched. The second rocket performed flawlessly, rising high into the heavens.

In Germany, the news related the progress of the war, on all fronts, was growing increasingly worse. In November of 1942, Rommel's Afrika Korps was beaten back by the British at El Alamein. American forces landed at Morocco and Algeria a few days later, forming a pincer against Rommel's remaining forces. In late November, Soviet forces launched a counterattack at Stalingrad. On February 2, 1943, Hitler received news that Field Marshal Friedrich von Paulus had surrendered the German Sixth Army to the Russians. Hitler searched for solutions that might change the fortunes of war for the Reich. He mulled over the rocket program in his mind. It is not altogether clear exactly what Hitler's attitude was toward the A-4 at this time. His approval of missile deployment, and also the behavior of others such as Speer, Degenkolb, and Himmler suggests that the rocket had become one of the highest priorities in armaments production.

In the summer of 1943, German forces launched an attack in Russia aimed at the Kursk salient, which the Germans intended to isolate. By trapping large numbers of Russian troops, Hitler hoped to regain the initiative. However, he was late in ordering the attack, and the Russian troops were dug in, having built heavily fortified defensive positions. It was Stalin who gained the initiative during the ensuing Russian counterattack, which turned the tide against Germany on the eastern front.

Early in July, Speer met with Degenkolb in Berlin. The Special A-4 Committee in its first days had completed a review of the industrial firms available for rocket production. Degenkolb had underestimated the complexity of putting the A-4 parts production into effect. Since only a few reputable companies were available, it was concluded that the commencement of A-4 production would be a daunting task. It would require a special decree from Hitler to speed A-4 production through more competent firms—firms that were already choked with other weapons manufacturing.

The scientists and engineers maintained legitimate doubts as to whether they could meet the excessive demands dictated by Nazi leadership. On June 5, 1943, during a national radio address, the Reich's propaganda minister Joseph Goebbels mentioned

Test Vehicle 41 (A-4/V41) is launched from P-10 (Test Stand X, just outside of the P-7 arena) on July 9, 1943. Immediately after ignition, the rocket canted horizontally and plowed into the pump building nestled against the outside wall of P-7. At the time of the explosion, engineers were preparing another rocket in the center of the arena. (NARA)

the coming of the wonder weapons indirectly. He asserted that soon Germany would strike back in retaliation, Germany would not sit idle while Allied bombers attacked German cities, and that the hour of vengeance would soon be at hand. The radio address was also heard in London. Goebbels, who had played an important role in helping the Nazis achieve and retain power by presenting the Nazi ideology to the German people in a favorable light, referred to the flying bomb and the A-4 as Vergeltungswaffe Eine and Zwei (revenge weapon 1 and 2), respectively. Henceforth, they would be known to the world as the V-1 and V-2.

The phone rang in Dornberger's flat early on the morning of July 7, 1943. On the other end of the line was Albert Speer. Minutes later, when the conversation concluded, Dornberger rang up von Braun and told him to get together some necessary things; they were going on a trip. Speer had summoned the men to Hitler's headquarters, the Wolfsschanze (Wolf's Lair), at Rastenburg in East Prussia. He told Dornberger to bring along with him the films of the successful test flight of October 3, 1942. Arriving early at the Peenemünde West airfield, Dornberger met von Braun. As a thick fog hung over the airfield, they loaded the films, a model of a proposed big firing bunker for the Channel coast, models of the rocket and select vehicles, and various drawings, manuals, and organizational plans all onto the waiting Heinkel He-111. The aircraft was piloted by Dr. Ernst Steinhoff, head of the A-4 Guidance Department. By late morning they had taken off. They were on their way to apprise the Führer—in person this time; to show the workings of the rocket.

They arrived early and were escorted by staff car to a guest house and told their appointment had been postponed until five o'clock that evening. Later in the afternoon, they started off an hour early to the restricted areas of the compound. They were escorted into the deepest reaches of the Führer's headquarters. The small cinema building was located in the middle of the compound. They set up their display of models and charts in the projection room and waited. The appointed time came, and Hitler was delayed, obviously attending to the details of his newly commenced offensive in the east. It was very late in the evening when, suddenly, the door swung open with the exclamation of "Der Führer!"[25]

With his hands crossed at his waist, Hitler shuffled into the small theater, followed by Albert Speer, Field Marshal Wilhelm Keitel, General Walter Buhle, Colonel General Alfred Jodl, and their aides. Dornberger was slightly taken aback when he saw Hitler's appearance. It had been almost two years since he had last seen Hitler, and the Führer seemed tired, his face full of lines, like an old man. After a short greeting with Dornberger and von Braun, he sat down next to Speer on the front row of seats. Dornberger composed himself and started into his prepared dissertation. After some introductory remarks, he motioned for the lights, and the room grew dark.

The projector flickered, and immediately onto the screen came the ascent of A-4/V4. Dornberger was, even now, moved when he saw the footage. He could only hope Hitler would also be moved, just as so many others had been upon seeing the incredible sight of the monstrous rocket slowly rising to the heavens. Von Braun commenced his commentary. The incredible scenes highlighted not only the flight but the breadth of the Peenemünde facilities, which Hitler had never seen. It went on to show an animation of how the rocket was tracked electronically, followed by scenes of Peenemünde prior to the war. At the end were the superimposed words "Wir haben es doch geschafft" (We made it, after all).

The end of the film flipped off the reel and then—silence.

Hitler was overwrought. Amazed by what he had just seen, he shifted back in his chair, gazing intently at the blank screen. The lights came up. Dornberger then broke the silence by entering into some explanations of the footage. Hitler jumped to his feet and strode over to Dornberger, listening very closely. He moved from the tables with the models and back again close to Dornberger, listening to the explanation of how the rocket could be deployed in combat. The two theories of operational deployment—a large concrete firing bunker or the use of mobile launching batteries—were explained. Dornberger talked about the formation of launching crews, their training program, and the expected capacity for monthly production of missiles. After referring Hitler to the charts and graphs, Dornberger finally stopped his presentation and waited for questions.

The Führer was like a changed man. Excitedly, he came over and shook Dornberger's hand. "Thank you. I am sorry that I did not believe in the success of your work from the beginning." Apparently, it was one of the rare occasions, if any, where Hitler apologized to anyone. Hitler continued, "If we had developed these rockets in 1939 we would have never needed to go to war. Europe and the world will be too small now to contain a war. With such weapons, humanity cannot endure it."

Dornberger could not believe his ears. Finally, once and for all, it looked like Hitler might take the rocket program seriously. However, Hitler showed greater interest in the model of the bunker than anything else. The model was explained again: how the rockets would be brought up, stored, and tested, then made ready for launching. Von Braun explained how they would be moved from the bunker, outside through giant sliding doors, only one minute before launching. Dornberger interrupted, saying he preferred the idea of mobile batteries, firing and then moving, thus avoiding the attacks from Allied airpower. It was like Hitler didn't hear Dornberger.

"Speer, come over here," Hitler summoned. "Wasn't it this same type of bunker that proved so useful to our submarines on the coast? We should build more than one, maybe two or three of these to protect our rockets!"

Hitler told Dornberger, in his opinion, the motorized rocket batteries would be spotted and attacked by enemy fighter aircraft. Dornberger respectfully disagreed, saying that after a rocket was launched, it would be extremely difficult for the enemy to spot the launching area. The firing site could be moved quickly, and only a handful of vehicles would be present at any point in the procedure. It did no good. Hitler was a bunker enthusiast.

Hitler then asked Dornberger if it were possible to increase the size of the warhead on the rocket. Dornberger explained that the rocket and it components were an integrated system, and the size of the warhead could not be increased without designing a totally new rocket.

"I want annihilation, complete annihilation!"

Dornberger replied, "When we started our work on the rocket, we were thinking in terms of long-range artillery. We were not thinking of an overall annihilating effect."

Hitler, in one of his abrupt mood changes, turned and angrily shouted, "I know *you* didn't think of it, but *I* did!"[26]

Dornberger stopped speaking. Field Marshal Keitel interrupted; changing the subject, he noted that Peenemünde should receive added antiaircraft defenses. Von Braun, also trying to calm the Führer, quickly pulled out several photos of craters made by impacting rockets. He explained to Hitler that the rocket would hit the ground at speed of over 1,000 meters per second and that the shattering force of the impact would multiply the destructive effect of the warhead.

"I don't accept this thesis," Hitler interrupted. "It seems to me that the sole consequence of that high-impact velocity is that you will need an extraordinarily sensitive fuse so that the warhead explodes at the precise instant of impact. Otherwise the warhead will bury itself in the ground, and the explosive will merely throw up a lot of dirt."

Some time later, von Braun conducted further research pertaining to the Führer's hypothesis; it turned out to be at least partly correct.[27]

Despite the warhead's apparent shortcomings, Hitler was still impressed. "I think we have something quite formidable here. The A-4 is a measure that can decide the war. What encouragement to the homefront when we attack England with it! This is the decisive weapon of the war; and what is more, it can be produced with relatively small resources. Speer, you push the rocket as hard as you can. Labor and materials must be fully provided," Hitler exclaimed.

The V-weapons programs benefited from Germany's misfortunes on the conventional battlefield. But Dornberger, after touting the advantages of the rocket for so many years, was now uneasy with the Führer's fervor and thought Hitler expected too much of the rocket. The propaganda surrounding the forthcoming Wunderwaffen, which were now being referred to as the Weapons of Retaliation, was giving rise to exaggerated hopes. These hopes sprang, in part, from Dornberger's relentless promotion of the missile. Dornberger now realized what was truly expected of the program, and it frightened him. Hitler and many others believed the secret weapons could cause a favorable change in the eventual outcome of the war. Peenemünde had developed a long-range weapon, which was just a weapon, not a weapon that could destroy all of Europe. While critical supplies of conventional armaments were needed to stem the Russian advance, Nazi leaders had ordered the startup of an expensive missile program.

When the briefing concluded, Hitler shook hands with von Braun and said, "Professor von Braun, I should like to congratulate you on your remarkable achievements." Professor? Unbeknownst to von Braun, Speer had secured the official title for the young rocket scientist a few weeks before, around the time when Dornberger had also been promoted to the rank of General. Saying good-bye, Hitler then left the room unceremoniously with his entourage.

After midnight, Speer entertained Dornberger and von Braun at the tea house of the headquarters. This was probably when the actual details of the meeting were worked out. Speer was the real driving force behind the production and deployment of the rocket. It was Speer who had convinced the Führer of the rocket's usefulness in light of the Luftwaffe's shortcomings. In early June, Speer secured the highest priority for rocket production—a priority that placed the V-2 above all other armaments production. The decision to deploy the V-2 had already been made; things were already underway. This meeting emphasized the rocket's importance in Speer's mind and sealed Hitler's confidence. The next morning the weather had cleared, and they flew back to Usedom. As the aircraft approached the peninsula, the vast extent of research establishment came into view. How impressive it was from this vantage point! Little did the men know it would be the last time they would see Peenemünde untouched by war.

3

Discovered

As early as 1939, British intelligence was aware of secret weapon trials on the north German coast near the Baltic. The anonymous Oslo Report received by British Intelligence mentioned several new weapon developments. The German tests focused on long-range weapons, but the precise location of these tests were not known. During the Peenemünde trials of A-4 Test Vehicle 2, the rocket center had been photographed for the first time on May 15, 1942. The RAF pilot, who was returning from a mission over Kiel, flew over Peenemünde and noticed an airfield and the strange elliptical works of P-7 below. Inquisitively, the pilot, Flight Lieutenant D. W. Steventon, turned on his camera. The photographs were turned over to photo interpreters in England. The strange earth works, heavy construction, and series of large buildings puzzled the interpreters. British intelligence analysts and scientists attempted to learn about the V-2 and to understand the nature of the German secret weapon program.

On the evening of April 15, 1943, Prime Minister Winston Churchill received an important memorandum. It read, "The Chiefs of Staff feel that you should be made aware of reports of German experiments with long-range rockets. The fact that five reports have been received since the end of 1942 indicates a foundation of fact even if the details are possibly inaccurate." The memo contained many inaccuracies; for example, it proposed that the Germans were developing a multistage rocket, capable of delivering 1,600 pounds of explosive over a range of 130 miles. Nonetheless, the bottom line was that the authors believed that a rocket weapon was probably in development. In the War Cabinet there was much skepticism about a long-range rocket, because scientific experts in Britain considered it impossible to build such a rocket. Not all members of the Chiefs of Staff believed the reports. Some felt the reports were disinformation intended to mislead British war planners. An aerial reconnaissance program was organized to cover essentially every square mile of German-held territory from the French coast at Cherbourg to the Belgian border. Everything within a radius of 130 miles from London would be photographed.[1]

The fragmented and conflicting rocket reports prevented the British Secret Intelligence Service (SIS) from establishing a concrete link to the Peenemünde establishment. On March 22, 1943, SIS received an important clue courtesy of two German generals captured by the British in North Africa. General Wilhelm Ritter von Thoma and General Ludwig Crüwell were overheard discussing Germany's rocket program while being held in London. They suspected their conversations were being secretly recorded and spoke softly to each other in the hope that no one could hear. Von Thoma spoke of a rocket test he had witnessed at Kummersdorf while in the company of Field Marshal von Brauchitsch. Von Thoma, who was privy to many technical details concerning German weaponry, spoke of how the rockets had been moved to the north and how the weapon should function. He went on to say that much talk had circulated recently among military commanders about the coming rocket offensive.

"The major in charge there (Dornberger) was full of hope. He said to me, 'wait until next year, the fun will start soon,'" whispered von Thoma.

The two generals commented that they had heard no "bangs" while in London, so something must have "gone wrong with those rockets."[2]

Spurred on by this plausible information, which confirmed the Nazis were intending to bombard London with a new weapon, British authorities ordered further investigation. Aerial reconnaissance photographs of Germany and France, taken in early 1943, revealed the huge research center at Peenemünde, supported by its own power station. The area contained many structures common in the production of explosives, but there were also towers, cranes, and elliptical and circular emplacements that could not be explained. Photograph interpreters in London, the Central Interpretation Unit (CIU), prepared the first Allied photographic reconnaissance report on Peenemünde on April 29, 1943. Even though the general appearance of the site suggested it was a manufacturing facility for explosives, the CIU interpreters all agreed that the photographs indicated the presence of test stands and launch pads.

On May 14, 1943, an RAF reconnaissance aircraft flying at an altitude of 26,000 feet over Peenemünde snapped images of the P-7 arena. A few days later, reconnaissance aircraft flew over the construction at Eperlecques. British photograph interpreters were startled by the huge dimensions of the site and two other locations nearby; all three were connected by suspicious railway lines. The site was deemed to be a probable rocket-launching site.

Over at the British Air Ministry, Dr. R. V. Jones, head of Air Intelligence at MI6 (Military Intelligence, Section 6), had been gathering pertinent information relating to possible German secret weapons. Jones diligently forwarded any information concerning the existence of such rockets to the War Cabinet Office. However, upon reviewing the information, the Chiefs decided for some reason to go outside of their in-house scientific department for further investigation of the potential threat. It was a man named Duncan Sandys, Undersecretary to the Ministry of Supply, who was charged with coordinating information about the secret weapons. Sandys was a tall man with wavy red hair and a dashing appearance. He was Winston Churchill's son-in-law and was suggested by the British Chiefs of Staff to head up the investigation under the auspices of the code name Bodyline. Churchill, delighted with the suggestion, made the appointment quickly.

In May of 1943, Sandys was preparing to submit his first report to the War Cabinet in London. He scoured over the available evidence, including the ominous and mysterious Oslo Report. He then looked to the photographic evidence and found intriguing anomalies on the photographs of Peenemünde. The photographs taken over the research center in May of 1942 revealed a new airfield surrounded by circular embankments and other strangely shaped buildings. There had been no further investigation, and the photographs were filed away. But now, in light of additional information, the photographs seemed very important. Sandys ordered more intensive aerial reconnaissance of Usedom.

He next reviewed certain intelligence reports that had arrived in London during the last six months—evidence collected by Dr. Jones. In 1942 a Danish chemist reported overhearing a confidential conversation about a long-range rocket. Varying reports about rockets and missiles had also come from informants in Sweden, Norway, and Poland. There were continuing bits of information filtering to the British from the foreign laborers at Peenemünde. The Polish slave workers housed at the Trassenheide settlement had managed to get word to Polish Home Army Intelligence of strange, torpedo-shaped objects seen at Peenemünde. These various reports, if considered separately, would have been ignored. They were tantamount to rumors; offering conflicting and superficial information. But Sandys saw a pattern in the reports. For example, one account given by Danish fishermen was about strange objects with "flaming tails" lighting up the night sky in the Baltic near Usedom. To Sandys, the reference to Peenemünde seemed more than a coincidence.[3]

When Sandys inquired with scientific advisors about the possibility of long-range rocket development in Germany, he was told that neither the British nor Americans nor the Russians had progressed to anything more significant than small, solid-fuel artillery rockets with limited range. They concluded it was highly improbable that the Germans had successfully developed such a weapon, and even if they had achieved a moderate amount of success, they did not possess the capability for mass production of a long-range rocket. Nevertheless, Sandys believed the potential threat was too great to be ignored. Something more ominous than conventional aerial bombardment lingered in the subconscious minds of every Allied war planner. In was known that in 1938 German scientists were the first to uncover the fundamentals of uranium fission upon which the atom bomb would be based. Information existed to indicate that Germany was conducting research in the field of atomic energy. It was not known how far they might have progressed or if this research had spilled over into the development of atomic weaponry. It was a contingency that was far too grave for Sandys to ignore.

The RAF photographic reconnaissance units consisted of five squadrons equipped with an assortment of Spitfire and Mosquito aircraft. These squadrons were primarily in Oxfordshire with one squadron in Leuchars, Scotland. One Mosquito squadron at San Severo, Italy, was also earmarked to begin searching for the rocket sites.[4]

V-2

The first photographic evidence of a rocket at Peenemünde came after a flight on June 12, 1943. Photograph interpreters at Medmenham reviewed the images taken that day but actually missed the appearance of a V-2 out in the open. Fortunately, copies of the photographs were sent over to R. V. Jones at the Air Ministry. Jones immediately picked out the faded but discernable silhouette of a rocket and informed Duncan Sandys of the discovery. Upon closer inspection, the photos revealed a 40-foot tower and, on it, a cylindrical object about 35 feet long, blunt at one end and tapered at the other. Sandys never credited Jones or the Air Ministry with the discovery, but the information appeared in his subsequent reports.[5]

After giving the mishmash of meager information appropriate consideration, Sandys wrote an interim report and presented it to the War Cabinet. He suggested in his dossier that Germany had been working on a long-range rocket for several years. He also suggested that not only rockets but also jet-propelled aircraft along with flying bombs were in development. He mentioned that available information did not indicate exactly how much overall progress had been made; however, it seemed likely that the German programs were in an advanced state. He concluded the most likely target for these weapons was the large city of London. Soon after submitting his report, Sandys was directed by the War Cabinet to intensify his search for more solid facts.

Ensuing reconnaissance sorties provided additional detail. Finally, a mission over Peenemünde on June 23, 1943, produced clear imagery of several rockets lying horizontally on long trailers near what was thought to be a launching emplacement. A vertical column was observed, which later proved to be a rocket readied for launching. Photo interpreters also found four small, tailless airplanes (V-1s), which were thought to be jet-propelled aircraft or airborne rocket torpedoes.[6]

Overall aerial espionage was increasing over the occupied territories. British authorities discovered unexplained construction sites springing up all over French coast. All of the mysterious structures were pointing at London. By sifting through the available data, the British Air Ministry's Scientific Intelligence Office and the British Secret Intelligence Service realized the evidence of a German missile program was now unmistakable, with launching sites probably planned for the Pas de Calais. The Bodyline investigation worked on the assumption that the range of the rocket might be close to 130 miles. The British scientific experts believed the weapon must be a multistage, solid-propellant rocket, since it was inconceivable (to them) that the Germans might have already advanced to liquid fuels. In determining the approximate size of the rocket from the details on the aerial reconnaissance photographs, the experts believed each one should have weighed between 60 and 70 tons, with a warhead weighing almost eight tons. Going on these assumptions, they began looking for some type of projection apparatus near the Channel coast, believing something so heavy needed some type of boosted assistance during firing. They would later find such installations; however, these were related to the employment of the Luftwaffe's steam-driven catapult for the V-1, not for the long-range rocket. The British did not hypothesize that the V-2 could be launched from either fixed or mobile platforms hauled by tractor or by railcar. If the

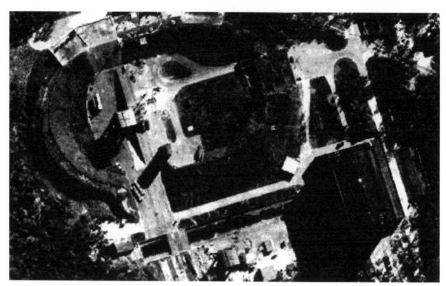

RAF reconnaissance mission over Peenemünde on June 23, 1943, produced clear imagery of several rockets lying horizontally on long trailers near what was thought to be a launching emplacement. One of the rockets may have been a complete 1:1 scale wooden mockup of a V-2 used for training purposes. It was painted white or possibly some other light color and was easily visible in the afternoon sunshine. (*NARA*)

rockets used some type of assisted launchers, then why were large tail fins observed on the rockets, which suggested the need for high stability during launch?

Based on the collection of accumulated evidence, Prime Minister Churchill called a full meeting of the War Cabinet Defense Committee members on the night of June 29, 1943. The meeting began at 10:00 PM in the underground headquarters of the War Cabinet in London. Churchill brought in his personal scientific advisor, Professor Frederick Lindemann (Lord Cherwell), along with R. V. Jones and Duncan Sandys. The meeting centered on the Churchill's desire to learn whether the rockets constituted a real threat or if the whole thing was some type of elaborate ruse. Duncan Sandys spoke first and presented his evidence supporting the theory of German development of long-range rockets. When Sandys finished, Churchill turned to his scientific advisor Professor Lindemann, who took an opposing view. He concluded that the shapes seen in the reconnaissance photographs were either dummy rockets intended to deceive Allied aircraft or ordinary barrage balloons. Professor Lindemann believed that the rumors of a 70-ton rocket were nonsense, which was absolutely correct. However, he never imagined the use of liquid propellant and concluded that the entire enterprise was a hoax designed to cover some other development such as a new type of large torpedo. After Professor Lindemann concluded, Churchill turned to Dr. Jones and said, "There is one last person I want to hear from on this matter."

Dr. Jones proceeded with a modicum of caution. Professor Lindemann was a colleague and a friend. It was Lindemann who recommended the young Jones for the position at the Air Ministry. Jones proceeded to tell Churchill that he agreed with the hypothesis of Duncan Sandys, although he felt the threat was not immediately imminent. He told Churchill he believed the evidence pointing to the existence of the German rocket was greater than any other assessment he had given the War Cabinet. For all intents and purposes, that ended the discussion. Churchill trusted the young Jones and immediately asked the group for proposed countermeasures.[7]

It was decided by the British War Cabinet that Peenemünde should be bombed at the earliest opportunity. The date would be late that summer, so that the bombers could fly the entire route under the cover of darkness, and would involve the whole of Bomber Command. Peenemünde presented a difficult challenge for the RAF bomber crews. The attack would require low-level pinpoint targeting with the bombers coming in at around 6,000 feet and dropping their bombs by moonlight. They also agreed that it was imperative to continue the reconnaissance missions along the northern French coast in the search for possible launching areas.

Back in Berlin, increased V-2 production demands had come from Karl Saur, Speer's deputy, during a meeting at the Ministry of Munitions. At the beginning of July 1943, Saur had invited 250 important managers and industrial heads, those who had received notification of pending rocket production contracts, to the important conference at the Ministry. Speer had already decided to resend Hitler's decree for increased armor production and to convert it into a decree for rocket production. The Ministry of Munitions stressed to the heads of industry the important details of Speer's plan to make rocket production the number-one priority.

It was announced that the A-4 Special Committee would add a new production facility in the Harz Mountains of central Germany. Committee members had discovered the existence of an underground mine* at the base of Kohnstein Mountain near Niedersachswerfen and Nordhausen in central Germany, which had been used during the early years of the war as a strategic storage depot for petroleum and chemicals. It was proposed to extend the tunnels even further into the mountain, creating a labyrinth for mass production of the V-2, deep underground where it would be safe from Allied air attacks. Degenkolb readily agreed to the plan; however, acquisition of the faculty was obstructed by Reich Marshal Hermann Göring, who was resentful of the rocket because of its new priority over Luftwaffe projects. Hitler settled the question, and the tunnels were soon emptied.

The new production facility would deliver 900 additional missiles per month, on top of the other 900 produced at Peenemünde, Wiener Neustadt, and Friedrichshafen. This new figure of 1,800 per month was immediately rounded up on paper to meet Hitler's goal of 2,000. Saur shocked the industrial contractors when he stated that this rate of production should be met by December 1943. Even if the production numbers could be met, there were not enough available chemicals to produce the propellants for even half of the rockets. Nevertheless, the industrial chiefs feigned approval, nodding their heads, although privately they scoffed at the excessive demands. Moreover,

*Wirtschaftliche Forschungsgemeinschaft (economic research partnership; Wifo)

Dornberger knew that the equipment needed to outfit the mobile launching batteries would not be ready for almost a year; what good would it do to have hundreds of rockets ready and no way to fuel or deploy them? The engineers and technicians at Peenemünde were upset when they heard that Degenkolb had approved the new, higher production figures and that Dornberger had agreed with the plan. The V-2 was not ready. During July 1943, four test rockets had been launched from P-7. All four were failures. The engineers wanted to proceed with realistic goals on a definitive timeline and felt the new plan was far too ambitious.[8]

From Berlin, Degenkolb continued his quest to speed rocket manufacturing. He brought in a man named Alben Sawatzki, who had been influential in working out the complex production details of the Henschel Tiger tank. Degenkolb dictated the aspirations of the "Degenkolb program" and entrusted Sawatzki with mass-production arrangements. He sent Sawatzki and three other production newcomers to Peenemünde to inspect the facility and become familiar with the rockets. Dornberger was anything but cooperative; after discovering their intentions, he immediately demanded they return to Berlin, with the exception of Sawatzki, who was allowed limited access to the pilot production facility. Although Dornberger had reason to be suspicious of Degenkolb's involvement at Peenemünde, his limited cooperation with industry experts slowed the progress toward production. The rocket was not designed for mass production and would need thousands of modifications to existing components. These deficiencies would severely delay the eventual deployment of the rocket.

The recent acceleration of Allied bombing attacks on German industry had prompted the Minister of Munitions to issue a plan whereby rocket production would be spread out and not concentrated in any one location. Peenemünde had already dispersed component manufacturing throughout hundreds of small firms all over Germany, and by subcontracting the multitude of specialized parts for the rocket, Speer hoped to minimize the effect of an air attack on Peenemünde or Friedrichshafen. However, this dispersal of manufacturing was causing turmoil in almost every quarter of manufacturing.

On the night of July 24, 1943, Sir Arthur Harris, chief of RAF Bomber Command, unleashed his bombers on the German city of Hamburg. Situated on the Elbe River with access to the Baltic Sea, the city of Hamburg was Germany's second-largest city. In a series of attacks ranging over nine days, more than 40,000 German civilians were killed. It was during these attacks that the RAF first made use of a diversionary tactic called "Window," the code name for the practice of releasing thousands of metal foil strips in an attempt to blind German radar stations. Caught off guard, the Luftwaffe fighter squadrons were fooled, and the British bombers got away relatively unscathed.

On July 25, 1943, Speer again traveled to the Wolf's Lair headquarters in Rastenburg to meet with Hitler about V-2 production. Upon his arrival, Speer found Hitler already in conference with his military commanders. Hitler was furious with Luftwaffe officials for what he perceived to be their complete ineptitude. The Luftwaffe's inability to thwart the British bombers or to offer any means of counterattack exasperated the Führer. It was Colonel Eckhardt Christian, Hitler's Luftwaffe aide, who took the full brunt of this admonishment. Colonel Christian tried to explain: during the attack on Hamburg, the British had released showers of Düppel (metal foil), which had jammed all German radar except Freya (which could not determine the altitude of approaching aircraft), resulting in mass confusion for the Luftwaffe. Hitler responded, "We can only stop the British attacks by striking back at English cities and civilians. You can only smash terror with counter-terror! They will only give in when we start killing their civilians."

It was immediately after this tirade that Hitler received Speer, who had come to him with a proposal for the Führer to grant the young Reichsminister sweeping powers in the propagation of V-2 manufacturing. Was Speer's timing a coincidence? Hitler readily agreed to sign the decree, which stated that the war in the west "required the utmost efficiency in missile output and that all measures should be taken to assure the immediate increase in V-2 production." Hitler felt the time had come for the weapon of revenge to be pressed into service.

With all of their attention focused on ironing out problems related to the missile itself, the engineers at Peenemünde had all but forgotten about the requirement for ground support equipment and specialized vehicles to outfit the rocket-launching batteries. The most important piece of equipment, the Meillerwagen erector trailer, had seen several fundamental design changes since 1942. However, no final production plans existed for the Meillerwagen or for the many other specialized vehicles that would be needed such as trailers, trucks, tankers, and electrical power suppliers. Dornberger picked the highly capable Klaus Riedel to oversee the job.[9]

Working with the Versuchskommando Nord, Riedel quickly brought in the necessary engineers to get the job done. Many of the battery vehicles were designed quickly on paper, notes and figures scribbled out, and then the request was sent to specialized German industrial firms for final design and production. It only took one month for the final production version of the Meillerwagen to be designed by the FX Meiller Corporation in Munich. By summer of 1943, a production run of about 200 trailers was underway at Rebstock, the underground Dernau-Marienthal complex.

The curious complex at Rebstock featured two unused railway tunnels through the mountain at Marienthal, near Bonn. The railway tunnels were converted into a fortified production center, protected from Allied bombing by the mountain. By August of 1943, almost all of the VKN personnel had relocated to the Dernau-Marienthal complex, along with a labor force of approximately 7,000 Russian and Italian prisoners. Production had begun on a wide range of battery vehicles, which, along with the Meillerwagen, included the Feuerleitpanzer (firing control vehicle), the Opel Blitz Sendewagen (which transmitted radio signal for engine cutoff), the Steyr Stromerzeugewagen (power supply vehicle), the Kabelwagen (large trailer that held the many electrical cables needed for the battery), and various other trailers and vehicles.

Following the July 7, 1943, meeting at Rastenburg, missile training commenced, and Dornberger was instructed to draft the necessary personnel. The Lehr und Versuchs

Batterie 444 (Training and Experimental Battery 444), commanded by Colonel Gerhard Stegmaier, was activated in mid-July. The unit was predominantly soldiers brought over from VKN. Dornberger gave Battery 444 the task of establishing a set of launch procedures for rockets in the field, training of additional firing crews, and evaluation of system performance and/or excessive difficulties during tactical operations.[10]

Differing opinions existed as to whether regular soldiers could be trained to fire the rockets. Presumptuously, the Peenemünde scientists and engineers believed that only qualified technicians such as themselves working in a self-contained environment like the Eperlecques bunker could successfully launch the rockets. They had been planning for this type of deployment, having plans and models already constructed. It was suggested the rockets would require an extensive support structure, such as the bunker at Eperlecques, which could store the necessary fuel and provide ready access to testing and maintenance equipment. All of the components were accounted for, including barracks to house the 250–300 personnel, antiaircraft batteries, and a way of transporting the missiles through a prepping stage, all the way to the moment of launching. Dornberger preferred mobile firing batteries manned by regular soldiers with specialized training. The military believed mobile launching crews, operating in a field environment, could avoid being detected by prowling Allied fighter aircraft. In the end, one battery was organized for the bunker and two others for mobile field operations. Dornberger's theory would eventually be proven correct.

In France, work continued on the massive bunker at Eperlecques. Laborers ceaselessly moved the 120,000 cubic meters of concrete required for the bunker. Inside the massive structure, the Germans intended to build a factory to produce liquid oxygen, thus avoiding the potential loss of oxygen through evaporation during transport. An extensive ventilation system was planned, including antitoxic gas filters where the fuels and explosive warheads were to be stored. There would also be lodging quarters for 250 soldiers and a multitude of antiaircraft batteries around the bunker.[11]

It was planned for rockets to be assembled in the northern portion of the bunker. This part of the site was accessible by two normal railway tracks, for supply of rockets and materials, linked to the track at Calais, Saint Omer. Between these two tracks, large trucks were to drive and park for access to the bunker. To the south were the towering halls at each end where the rockets were to be assembled and checked. In the forward galleries, the rockets were to be lifted up into a vertical position. They would then be fueled, the warheads attached, and lastly, the detonators fixed in. The five compressors to produce liquid oxygen were located between the two halls. The rockets would be moved to either end of the building through the 18-meter-high pivoting doors. They would exit through the south face of the bunker via two corridors. On the exterior, remote tracks would move the rockets to the firing pads. There would be no doors where the rockets passed through to the outside. A perpendicular door would be installed at the back of the exit corridor, away from the gases, to minimize the blast of launching rockets. Chicanes in the concrete on the side walls of this corridor were designed to buffer the shock wave as the rocket was being launched.

V-2

It would require 4.7 tons of liquid oxygen to launch one rocket. The German specialists estimated a loss rate of one percent per hour due to evaporation in storage and allowed for an additional five tons per rocket. This indicated a projected delay of more than four days between liquid oxygen manufacture and rocket launching, which brought the required amount to around 14.4 tons per rocket. Hitler fancifully decreed that the bunker should start by firing 144 rockets each day. This schedule would therefore have required about 2,073 tons of liquid oxygen per day. However, at the beginning of 1943, the total German production, including liquid oxygen plants in the occupied territories, would only allow for approximately 200 tons per day. On March 3, 1943, a letter was written in which the Germans were planning 23 production compressors, of which five would be at Watten. All 23 compressors together would have had a capacity of 84,000 tons per year, which added to Peenemünde, would have given 102,000 tons per year, or a rate of fire of about 19 missiles per day.[12]

For several days, the summer sun at its zenith had been beating down on the parched, sandy coastline of Usedom. On the morning of August 17, 1943, Dornberger summoned his managers and department heads to his Peenemünde office. For several hours the men met in a closed-door conference; they were not to be disturbed. Assembled were Colonel Stegmaier; Councilor Schubert, Chief of Production Works; Chief Engineer Rudolph; Engineer Stahlknecht; and directors Steinhoff, Thiel, and von Braun. The passionate discussions that followed centered on the difficulties presented by the Degenkolb schedule. The unusually oppressive heat added to the tensions that day as arguments ensued between the production planners, who believed production blueprints were excessively slow in coming from the development section, and the development engineers, who believed the exacting tolerances of the rocket components were too complex for industry to manufacture. At one point in the meeting, a dejected Dr. Thiel even threatened to quit Peenemünde and take a job at a university. Dornberger somehow managed to calm the group and convinced his colleagues that every problem had a legitimate solution. He would see to it that each of their concerns were addressed and urged them to keep a positive attitude. Dornberger pointed out that the men had plenty to be excited about. Peenemünde was experiencing a huge influx of support, with over 1,200 new workers arriving just in the past few weeks. Von Braun made some useful suggestions, and soon the group was discussing ways of lessening their impediments. Later that evening the men spoke cheerfully following dinner.[13]

General Dornberger retired to his quarters around 11:30 PM. Tired from the heat and the emotions of the day, he was about to drop onto his bed when he heard the howling of an early warning alarm signaling a forthcoming British air raid. He did not pay much attention to it. It had become common for the bombers to form up over the Baltic before they flew south to attack Berlin. Blacked-out Peenemünde, which was now bristling with newly installed antiaircraft defenses (88-millimeter flak guns ringing the tip of the peninsula), would simply let the bombers fly on. The research establishment did not want to draw attention, and the guns were to remain silent unless

Peenemünde itself was under attack. Dornberger phoned the air defense headquarters and was told, "Bomber formations are massing north of Rügen, destination unknown at this time." Even though Dornberger had been warned by officials in Berlin that Peenemünde could be attacked any day, he was asleep almost as soon as his head touched the pillow.[14]

Half a mile away, in the area of the Siedlung, or worker's settlement, 4,000 German technicians and their families slept quietly, including the family of Dr. Walter Thiel. Always early to bed and early to rise, Dr. Thiel and his family had gone to bed around 10:00 PM that evening. Immediately south of the settlement was the POW camp where 500 Russian workers were sleeping. Further south were the huts of the Trassenheide camp, where the 600 foreign slave laborers were held. Secured by SS guards with machine guns, the Trassenheide camp was surrounded by barbed wire and guard dogs.

Peenemünde's peaceful existence was about to end. On the night of August 17, 1943, RAF Bomber Command dispatched hundreds of bombers against the German research establishment. Code-named Operation Hydra, the attack, using the most sophisticated bombing techniques available at the time, commenced with a full moon and no cloud cover over the Peenemünde. It was a huge operation and included 596 bombers of varying types, including Short Stirlings, Avro Lancasters, Handley-Page Halifaxes, and de Havilland Mosquitoes. The route to Usedom crossed over the North Sea and then over the narrow neck of Jutland before crossing Zeeland. A short time over the Baltic brought the bombers to the Arkona Peninsula. It was from here they began their approach to the target. They had specific orders to bomb the workshops of the development center, the large factory, and to kill the scientists and technicians while they slept. The raid depended on the preliminary pathfinders to mark the aiming points, with time, distance, and coordination provided by a master bomber. Early in the operation, a squadron of fast-moving Mosquitoes dropped Window. Thousands of foil strips fluttered through the air, blinding German radar. The Mosquitoes then flew on to Berlin in an attempt to draw away the German night fighter aircraft. The scheme worked as anticipated and the pathfinder aircraft then dropped their flares over a sleeping Peenemünde shortly after midnight. Minutes later, the first wave of bombers started their runs over the research establishment. Most of the bombers reached Usedom with little interference prior to dropping their bomb loads. Unfortunately, the first wave of bombs fell too far to the south, killing more than 700 at the Trassenheide labor camp. The master bomber directed the ensuing two waves over the German housing and research areas. In the end, the RAF lost 40 aircraft and 290 young men, most in the latter waves, after German defense forces had recovered.[15]

Dornberger was shaken from his slumber by a tremendous detonation. He found his bed sheets covered in broken glass from the shattered window panes of his room. He could hear the antiaircraft guns raging in the distance and the steady drone of aircraft engines overhead. Reaching over, he grasped the telephone next to his bed and called the concrete command shelter, only to hear a busy tone. Jumping out of bed, he fran-

V-2

tically began looking for his boots, which were nowhere to be found. He put his coat on over his pajamas, found his slippers, and walked gingerly over the broken glass into the street. The scene that greeted him was unbelievable. Looking to the north, he could see the antiaircraft guns flash while the searchlights traced the night sky. The haze from the smoke canisters drifted through the air, accented by the radiant moon and the brilliant RAF marker flares. The concussion of the exploding bombs mixed with the sudden flashes and deafening roar stopped Dornberger in his tracks for a second. He turned his head left and then right; fires were burning everywhere it seemed. He raced across the road toward the command shelter.

Von Braun, his blond hair white with ashes, had already made it to the concrete shelter. Standing outside, he continually glanced skyward and then toward the bachelor's quarters, waiting to see who would emerge from the mist. Dornberger ran up and grabbed his arm just as an incendiary bomb exploded above their heads. The two men hurled themselves into the shelter. The room was filled with people in stunned silence. Dornberger looked for moment at the helpless faces and then quickly picked up the telephone. The air raid warden on the other end of the line confirmed that Peenemünde was in the middle of an all-out air assault. Dornberger thought for a second and then began issuing orders. He split the men into groups and ordered them out into the inferno to save the all-important documents. Von Braun took a group of men with him and made his way through the conflagration to the construction bureau. Putting out fires and gathering the thousands of papers representing over a decade of work, the men hurriedly made their way back to safety. Dornberger took another group in a different direction toward the measurement house, assembly workshop, and component workshop. After extinguishing small fires, checking, and securing the buildings, Dornberger noticed the guesthouse was burning. He rushed into the room, where he had been sleeping only an hour before, to retrieve some of his family papers and personal possessions. Minutes later the fire consumed the room.[16]

Dornberger and von Braun may have been satisfied that for the moment, the documents had been salvaged, but what they didn't know was that Dr. Thiel had already perished, along with his entire family, when their little wooden cottage took a direct hit from a falling bomb. In the settlement, hundreds of technicians and their families were sent scurrying for cover as the thousand-pounders rained down around their homes. Some of the initial target markers fell more than two miles from their intended target. This was fortunate for the rocket development center, but proved disastrous for the slave workers in the Trassenheide camp. A heavy concentration of incendiary bombs fell directly on their wooden huts, setting on fire those who were not killed by the explosions. In a panic, they ran to the fence surrounding the camp, begging the SS guards to let them free. They did not. The secret reports the British had been receiving from Trassenheide ceased after that night.

When the third wave of bombers arrived over Peenemünde, it was very hard to see the target markers or observe the effects of the raid because of the many fires and huge veil of towering smoke rising high into the air. To make matters worse, British aircraft began bursting into flames as the antiaircraft gunners found the range, and the Luftwaffe night fighters, which had been drawn away by the feint at Berlin, had now returned and were slashing through the bomber formations. Some 50 minutes after the raid had begun, the RAF bombers turned for home.

The following morning, Dornberger surveyed the damage. At first glace, the destruction appeared to be overwhelming. All along the seashore there were piles of debris and rubble where pristine homes and workshops once stood. Rail and road networks were pockmarked with giant craters, and hardly pane of window glass was left unshattered, apart from the windows in the development works, which were reinforced with tiny strands of wire embedded in the glass. The damage to the research center was considerable. However, it was not as extensive as the British had hoped. Dornberger estimated that it would take as little as six weeks to recover from the attack. Not only were all the important documents saved, the most critical installations such as the test stands, wind tunnel, and measurement house were relatively untouched. Hall F-1, the pilot production facility, had received several hits but was not severely damaged. The huge facility was almost complete but not operational at the time of the attack. Inside were several completed V-2 test rockets in storage.

However, the settlement where the thousands of German technicians and their families lived was all but wiped out. Four days after the raid, a mass funeral was held near the railway tracks north of the smoldering Trassenheide camp. A long communal grave was dug for 735 victims, the majority of which were the POWs from the Russian camp and foreign slave laborers from Trassenheide. Among the bodies interred were several dozen members of the VKN, whose quarters were very near the Trassenheide camp. Civilians from the housing estate would be buried in separate graves. A total of 178 German technicians had been killed; among these was the talented Dr. Walter Thiel. With his intimate knowledge of rocket engine design, Thiel would be the hardest to replace. After the attack, Dornberger knew things would never be quite the same. The significance of the raid was not the damage inflicted on the research center, it was the revelation to the Nazi leadership that the Allies recognized the importance of Peenemünde. Consequently, the raid was responsible for drastic and sweeping changes in the program; changes that at the time, however, Dornberger may not have fully anticipated.

Following the raid, General Dornberger was not the only person inspecting the damage to the facility. Upon hearing of the attack, Reichsführer SS Himmler immediately dispatched his head of security, SS Obergruppenführer Ernst Kaltenbrunner, to Peenemünde. Himmler realized this was an opportunity to exercise SS influence over the rocket program. He had long sought to wrestle control from the Army and asked Kaltenbrunner to find an excuse for the SS to become involved. Himmler received Kaltenbrunner's report while at Hochwald, the SS headquarters near Grossgarten and Schwenten, 25 kilometers north of the Wolfsschanze. On August 19, 1943, he made the short drive to the Führer's headquarters and once again asked for control of the rocket program. He proposed that V-2 testing and training be moved to occupied

Poland under SS supervision at a place called Heidelager, near the small village of Blizna, and also suggested that if the missiles were constructed entirely by slave laborers, deep underground, greater secrecy could be maintained over the project. This is ironic, since it had been the Polish and Luxembourg workers at Trassenheide, under SS control, who were able to get word to London about the missiles at Peenemünde. Nevertheless, in light of the bombing raid, Himmler's ideas seemed prudent to Hitler, and he approved of the plan. The Reichsführer SS, by controlling all concentration camp labor, had gained authority over a good portion of the rocket program. The SS was trying to draw every institution in Germany into its tentacles.[17]

On Hitler's orders, Armaments Ministry officials Speer and Saur traveled to SS headquarters at Hochwald to meet with Himmler the following morning. Himmler announced that as the newly appointed Minister of the Interior, he had been charged by the Führer to look into the best solution for protecting the rocket production program. A few days later, Speer received a letter from Himmler stating, in effect, that the SS was taking total control over rocket production. Speer was forlorn, but there was little he could do about Himmler's coup. Allied bombing had placed heavy restrictions on Germany's abilities to expand its war production. In spite of Hitler's top-priority directive for Peenemünde, Speer encountered extreme difficulties in his efforts to manipulate industry to produce the rocket. Speer was losing power to Himmler.

Himmler quickly named as his deputy 42-year-old SS Brigadeführer Hans Kammler, an engineer who had been one of the primary designers of the secret extermination camps. Dr. Kammler had become a rising star in the ranks of the SS since first joining in 1941. In 1942, he was appointed head of Division C, the SS construction wing of the Economic Administrative Main Office. Following the Warsaw ghetto uprising in 1943, Himmler assigned him to oversee the demolition of the ghetto in retaliation. Kammler came from a middle-class family, received his Doctorate of Engineering from the university, became prominent because of his work in construction, and advanced through the SS rapidly into other fields for which he really had no training. Upon first impression, he seemed to be every bit the ideal, handsome, Nordic figure—the superlative SS soldier as one might have seen on a propaganda poster of the day. When Dornberger first met Kammler, he described him as "a slim figure, of perfect build, with bronzed clear-cut features. He looked like some hero of the Renaissance." However, those who were impressed by Kammler's initial charisma would soon realize that his vigor, objective coolness, and determination scarcely cloaked extreme arrogance and brutality. Speer said of Kammler, "I discovered him to be a cold, ruthless schemer, a fanatic in the pursuit of a goal, and as carefully calculating as he was unscrupulous."[18]

Reichsführer SS Himmler was making every effort to ensure that Germany could survive. He demanded that factories for the production of war materials be built in natural caves and underground tunnels immune to enemy bombing. His goal was to have the greatest number possible of the new underground factories completed by 1944. He instructed Kammler to hollow out workshops and factory space for these uniquely bombproof work sites. The urgency with which the underground rocket program was pressed forward was fostered by Himmler's mistaken belief that the threat of a missile assault on London might force the west to the negotiating table.[19]

In late August of 1943, the new production plan proposed by Speer involved the intermingling of several factories, with the Mittelwerk (Central Works) operating as a final assembly point for the rockets, while other plants such as the Zeppelin plant in Friedrichshaven and Henschel's Rax-Werke in Weiner Neustadt (the Southern Works) and a potential new plant in Latvia (the Eastern Works) manufacturing subassemblies. However, after Kammler became involved, this arrangement did not last long. Less than a week after Peenemünde had been bombed, Kammler announced plans for the Mittelwerk (Central Works), which was the new name of the production plant at Niedersachswerfen. Kammler proposed three new projects to the rocket experts. These included the main factory for manufacturing in the Kohnstein Mountain, a development plant in Austria (to replace Peenemünde), and a new firing range for testing of rockets in Poland. Dr. Kurt Kettler, who had been director of the Borsig Lokomotiv Werke G.m.b.H. in Berlin, was named general director of the new corporation, Mittelwerk G.m.b.H. (Central Works, Ltd.). The Mittelwerk would take over all rocket assembly, with specialized component manufacturing distributed throughout German industry. Kammler also appointed all of the management personnel for the newly created Mittelwerk "corporation," who would work closely with the SS commandant of Camp Dora in the exploitation of the slave laborers. Degenkolb put the first part of the plan in motion by jump-starting the construction of the camp needed to accommodate 30,000 forced laborers near the underground factory. On August 28, 1943, the first truckloads of prisoners arrived at Niedersachswerfen from the Buchenwald concentration camp to begin the hard labor of expanding the Wifo tunnel system and to create within the tunnels the Buchenwald subcamp known as "Dora," which was later independently known as KZ Mittelbau. On September 2, 1943, more than 1,200 additional slave laborers, mostly French, Polish, and Russian, arrived at Dora. Under the direction of Kammler, the underground factories were built with astonishing speed but at the expenditure of thousands of slave laborers, many of them Jews.[20]

The tunnel system of the Mittelwerk had initially been started in 1934 as a gypsum mine. In October 1940, the Armaments Ministry in Berlin approved expansion of the Wifo site. Throughout the fall of 1943, the Dora prisoners endured physically demanding work, struggling in horrendous inhuman conditions to enlarge and expand the tunnels. Prisoners drilled and blasted away thousands of tons of rock. They worked atop 30-foot scaffolds, using picks to enlarge the tunnels. From time to time, a prisoner would become too weak to continue, fall to his death from the scaffolding, and be replaced by another. They built rickety, temporary narrow-gauge tracks to support the multi-ton loads of rock that were extracted from the caves. If the "skips," or small rail cars, full of rock fell off these tracks, which happened frequently, prisoners were kicked, whipped, and beaten until they reloaded the cars. They were forced to eat and sleep within the tunnels they were digging. Thousands of workers were crammed into stinking, lice-infested bunks stacked four high in the first few south-side cross tunnels at the mouth of Tunnel A, in an atmosphere thick with gypsum dust and fumes from the blasting work, which continued 24 hours a day. They had no running water or sanitary facilities, and dysentery, typhus, and tuberculosis, along with starvation, were constant causes of suffering and death for these unfortunate people. All of the manufacturing equipment from Peenemünde had to be transferred and installed in the tunnels. This was done by hand, using hand carts, block and tackle, huge skids pulled by teams of prisoners, and the temporary narrow-gauge rail lines. By the end of 1943, 46 cross tunnels existed, and each of the main tunnels was wide enough to allow for rail traffic directly into the heart of the mountain. Trucks carrying piles of dead bodies left every other day for the crematorium ovens at Buchenwald, prompting the prisoners there to observe, "The only thing that Dora produces is corpses."[21]

Hitler ordered the construction of sufficient numbers of liquid oxygen manufacturing facilities to support the approaching rocket offensive. He demanded that the plants be established underground, protected from the incessant Allied bombing. Liquid oxygen plants were located at Lehesten, south of Saalfeld; Raderach near Friedrichshafen, which was the first to open; Redl-Zipf in Austria; and Wittringen on the Saar. Late in 1943, prisoner labor was used to carve the Lehesten facility out of a quarry site about 128 kilometers southeast of Mittelwerk in southern Thuringia. The underground tunnels at Lehesten offered an ideal location for a liquid oxygen plant. Built into the side of the quarry were two concrete static test stands, about 76 feet apart, used for final test firing and calibration of V-2 engines, pumps, and valves before they were shipped to the Mittelwerk for final assembly. The Lehesten plant would produce about eight metric tons of liquid oxygen each hour, closing down for four days each month to thaw the equipment. There was also a similar facility located at Redl-Zipf in Austria. Lehesten had its own small concentration camp, called Laura, located in the village of Schmiedebach. Laura was also a subcamp of Buchenwald.[22]

With the mission against Peenemünde completed, the RAF turned its attention to the French coastline. The Germans had proposed firing rockets against cities in southwestern England as well as the large population center of London. For the battering of London, the Germans needed catapult installations for the V-1s and either bunkers for fixed V-2 operations or supply depots for mobile operations in the Pas de Calais area. For the other targets such as Aldershot, Winchester, Plymouth, and Southampton, the facilities would need to be located in Normandy. It is unknown exactly how many launch sites were planned, but it may have been as many as 40 to 50 V-2 sites and 60 to 100 V-1 sites, all in a wide semicircle from Dunkirk to Cherbourg.[23]

Given the perceived potential threat posed by the V-weapons, the whole of RAF Bomber Command and an appreciable number of heavy bomber units from America's Eighth Air Force were now tasked with destroying the German rocket sites. Almost half of all Allied photoreconnaissance was now devoted to the rocket threat.[24]

On May 16, 1943, the first aerial photographs of the Blockhaus at Eperlecques had been taken by RAF reconnaissance aircraft. As the RAF planned its night raid on

The V-2 rocket bunker at Eperlecques near the Channel coast. Project Kraftwerk Nordwest (Power Station Northwest) KNW was constructed by the Germans in 1943 to accommodate V-2 reception and storage, preparation of rockets for launch in a sheltered and controlled manner, on-site production of liquid oxygen, and launch control with two firing pads. The bunker drew the attention of Allied war planners and was attacked by Allied bombers on August 27, 1943. Following the raid, the site was deemed unsuitable for launching rockets; however, construction continued in attempt to finish the liquid oxygen manufacturing portion of the bunker. It was finally knocked out in 1944 after 20 Allied raids, some using the enormous Tallboy bombs. Canadian forces captured the site on September 6, 1944. (NARA)

Peenemünde, photographic over flights at Eperlecques showed that a number of rail lines and huge underground bunkers were being constructed near Watten. Intelligence sources reported that as many as 6,000 construction workers were seen working at the site. Subsequent Allied photographs at the end of August revealed construction workers were in the process of pouring thousands of cubic meters of concrete on the site. On August 27, ten days after the raid on Peenemünde, 187 Flying Fortresses of the Eighth Army Air Force, escorted by 147 P-47 Thunderbolts, attacked the bunker in the late evening. After crossing the coast, the B-17s were attacked by German fighters, which took advantage of the lack of coordination between the bombers. Despite the deadly accurate antiaircraft fire and attacks from German fighters, the attack continued for about an hour with a total of 366 bombs being dropped. The 2,000-pound bombs devastated the huge site, especially the northern section where large quantities of concrete had just been poured, leaving a hardened, twisted mass. At the time of the attack, workers had completed more than a third of the total construction. Some of the forced laborers died in the attack, while others took the opportunity to escape. During the raid, the Americans suffered the loss of four B-17s, with 98 others being damaged, along with one P-47 shot down. Upon landing in England, the crew of one B-17 counted more than 200 flak holes in their aircraft. The American aircrews claimed a dozen German fighters shot down.

The daylight attack of August 27, 1943, was the first time the German V-weapon sites had been targeted by the U.S. Eighth Air Force. For good measure, from August 30 to September 7, the complex was attacked four more times in smaller raids using medium and heavy bombers. Officials from Organization Todt soon deemed the northern section irremediably lost. At the time, German officials still believed the Pas de Calais area in northern France would be the eventual launching area for V-2 rockets, and even if the bunker was not used to prepare and fire rockets, liquid oxygen would still be needed to supply any potential mobile field operations. During the months of September and October, Organization Todt investigated the Blockhaus and decided to complete the southern portion of the bunker for liquid oxygen manufacturing. One of Todt's top engineers, Werner Flos, had an idea to continue the construction using a technique that would protect the site from aerial bombardment during construction. He suggested building the roof first and raising it up from the ground. In November, the southern portion was cleared and new work started by pouring a concrete roof five meters thick, in sections. The roof would be raised gradually by using a series of giant hydraulic jacks and blocks. The exterior and interior walls would be built underneath. This roof would protect the construction taking place below it. The strata of each concrete layer were cast, and each time, the roof was raised to build the outside wall. In this manner the building was raised to 28 meters. Even though the new building was to be a liquid oxygen production facility only, features for the movement and launching of rockets were still incorporated in its new construction. This meant that the Germans might have held out the possibility of launching a limited amount of rockets from the site.[25]

Meanwhile, workers at Peenemünde had been focused on cleaning up and restoring operations. In an effort to deceive Allied reconnaissance, many of the damaged structures were left exactly as they had been following the raid. Only the most essential buildings were repaired, and these were rebuilt in a manner that made them appear seemingly ruined when viewed from the air. Hereafter, there would be no attempt to manufacture rockets at Peenemünde. All manufacturing would be transferred to the Mittelwerk in central Germany. The supersonic wind tunnel would be transferred to Kochel in the Bavarian Alps. A V-2 was launched from Test Stand VII on October 6, 1943—the first since the air raid on Peenemünde seven weeks earlier.

In an effort to maintain Army control over the rocket program and ensure that Dornberger was not cast aside by the new influences of the SS, General Fromm had removed Dornberger from his position at Army Ordnance and on September 4, 1943, named him as his direct subordinate with a new title, Beauftragter zur besonderen Verwendung Heer (Army Commissioner for Special Tasks, or BzbV Heer). He also was given control of the world's first ballistic missile command, Höhere Artillerie Kommandeur (Motorisiert) 191 (Superior Artillery Commander, Motorized, 191, or HARKO 191 for short). He maintained control over development of the operational missile; however, Dornberger's specific job was to be the recruitment and training of the operational rocket batteries and to assume operational command over military deployment of the V-2 when the time came. He was also charged with establishing a logistical supply system for all required materials in the operational areas. Dornberger's operational command was affirmed by Hitler on October 4, 1943, when he once again visited the Führer at his headquarters in East Prussia. However, it wasn't long before Hitler would dilute this decision also, causing, once again, great frustration for Dornberger.[26]

British intelligence estimated that the damage inflicted upon the Eperlecques large site by the Americans would delay construction about three months. A few days after the raid, German engineers suggested a rock quarry at Wizernes as an alternate prospect for V-2 bunker operations. The site was 12 kilometers southeast of Eperlecques, closer to Saint Omer near the Channel coast. The site was originally intended as an underground storage depot in support of the Eperlecques facility, but now new plans were drawn up. Wizernes would become the most ambitious V-2 launching bunker of all. During a meeting with Hitler on September 30, 1943, a man named Xavier Dorsch, chief engineer with Organization Todt, proposed the site to Hitler, Himmler, Speer, and Dornberger. During construction he suggested using a variation of the Verbunkerung technique in which a million tons of concrete would be used to construct an impenetrable dome on the hillside overlooking the quarry and a series of connecting tunnels; about seven kilometers of underground galleries would be excavated beneath the dome and hillside. The dimensions of the dome would be gigantic: 71 meters in diameter and five meters thick, weighing 55,000 tons. Hitler, as usual, was ecstatic about the grandiose plan. The Wizernes domed-bunker, also known as the "Cupola," would receive the code name of Bauvorhaben 21—Schotterwerk Nord West (Building Project 21—Gravel Quarry North West).

In October of 1943, Organization Todt began initial construction at Wizernes. On paper, it was one of the most imposing structures related to the V-2 program. Organization Todt entrusted the work to large German construction companies, and soon an enormous amount of hardware arrived such as earth-moving and drilling equipment along with concrete batching and mixing plants. More than 1,300 forced laborers descended on the site, working day and night, badly nourished and abused by the German guards. The foremen and the skilled workers were German; the hard laborers were forced workers, young Frenchmen forced into the STO, and captive Soviet POWs.

The building plans of the Cupola were titanic. The galleries were dug in the chalky plate for storage of the rockets, fuel supply, a liquid oxygen manufacturing facility, housing for the garrison, and the generators for electricity. There were to be paths leading from outside through rocket-high doors into the interior assembly area. Inside, the rockets could be serviced and assembled safely, shielded from Allied bombing by the massive dome and the chalk hill as well. A small railway supply tunnel would lead to all of the underground workings and to a large octagonal chamber. This chamber was 41 meters in diameter and seven stories high for fueling and prepping of the rockets. Another chamber for preparation of the rockets would be located directly under the dome. When ready to fire, a motorized platform would move through the giant tunnels, nicknamed "Gustav" and "Gretchen," through the five-foot-thick, 55-foot-high, solid-steel exterior doors to the outside, where the rockets would be launched quickly. A specially trained battery of V-2 troops would run the operations of the bunker.

It is interesting to note that the Wizernes was built to incorporate the features of the Regenwurmlager (earthworm camp) concept. Along with the building of the dome complex for remotely firing V-2s, the Germans built predefined locations for V-2 mobile launching units, called Regenwurm Stellungen, or earthworm positions. There were between 30 and 50 of these sites scattered across the countryside around Wizernes. Many of these concrete launching pads can still be found today. The idea involved moving prepared V-2s from the Wizernes complex on trailers, which would be parked in a planned (but never constructed) tunnel system. By November 1943, construction had started on the dome, along with the tunnel excavation at the base of the quarry. Although aware of the existence of an abnormal building site close to Saint Omer, the Allies were slow in targeting the bunker for air raids.

One very unique structure related to the Wizernes project was located in the small French hamlet of Roquetoire, about eight kilometers away from Wizernes and southeast from Saint Omer. Inside of this fortress-like concrete structure would be an ultramodern, ground-based, remote radio guidance beam establishment for V-2 operations called the Leitstrahlstellung (beam position). This apparatus was conceived and built for guide-beam flight correction of the V-2 rocket during launch. Roquetoire received the code name of Umspannwerk C (Transformer Station C). The bunker would have provided protection for the vehicles and supplies of the crews operating the guide-beam equipment. It would have been a safe place for storage and repair, ensuring ease of deployment during operations. The beam system was so secret that only the planners and personnel at the two German facilities knew of its existence. Advances by the Allies in the use of jammers to break the German radio signals could have easily countered the guidance signal if it had been discovered. Even though the Leitstrahl radio-beam operation would last only about a minute during a V-2 launch, and any jamming attempts would have required a known frequency of the beam, the Germans kept all information about this device top secret. In fact, the Allies never discovered this system until after the war. The Roquetoire bunker would never become operational, but the Leitstrahl system would later be adapted for mobile firing operations.

Eperlecques was the first of seven large construction sites related to the V-weapons. In July 1943, Duncan Sandys suggested that the RAF should target the site; however, the British Chiefs of Staff were aware that the Americans were already planning daylight missions against the Blockhaus and were satisfied to let U.S. Army Air Forces (USAAF) carry out the raids. Meanwhile, British Intelligence discovered even more mammoth construction projects at Löttinghen and Wizernes, at Mimoyecques and Siracourt in September, and shortly thereafter at Sottevast and Martinvast on the Cherbourg peninsula. These bunkers had steel-reinforced concrete walls up to 30 feet thick and were large enough to house whole divisions.[27]

In addition to the large construction sites, the RAF detected hundreds of smaller works all over Pas de Calais and Normandy. Many of these locations featured some type of curious launching system consisting of two inclined rails, almost 300 feet long, resting on a metal latticework anchored with a concrete emplacement. Near the end of

October 1943, aerial reconnaissance revealed that these "ski sites" were installations for launching some sort of pilotless aircraft (V-1). At this point almost two dozen such sites had been discovered; all of the ramps seemed to be pointing at London.[28]

On October 31, 1943, the Allies discovered additional construction activity in the area of the Cherbourg peninsula. A massive site near Sottevast, a new giant rocket bunker called Reservelager West (Reserve Site West) and another site at Brecourt were each photographed extensively by Allied reconnaissance. It was not immediately known (but discovered later) that the two massive German works projects in the Cherbourg peninsula were intended for the launching of V-weapons toward England. The bunker at Sottevast was built using the graduated Verbunkerung technique and was intended for operations of the V-2 in much the same way as the Eperlecques site, in which testing, fueling, and servicing would be carried out in a hardened, bombproof shelter. Limited construction work commenced in October 1943 and continued through the spring of 1944, even under the constant threat of Allied bombardment, which began in February of 1944. The unfinished works were captured by advancing Allied troops in June of 1944.

After the initial investigation by Duncan Sandys and the Bodyline committee, the task of locating, photographing, and assessing new targets was so overwhelming that it had to be transferred to the larger intelligence departments of the Air Ministry. The Joint Intelligence Subcommittee of the British Chiefs of Staff replaced the code name Bodyline with the new name Crossbow. The War Cabinet received the first Crossbow assessments on November 29, 1943, and again ordered substantial aerial reconnaissance and unfaltering bombing attacks on the suspected launch sites in northern France. Operation CV, the Allied aerial offensive against the V-weapons, was the most extraordinary bombing offensive of the whole war. The menace presented by a potential German bombardment of its southern cities caused the British to divert its bombers from attacks on German industry and slowed interdiction bombing prior to Operation Overlord, the Allied D-Day invasion of France. There were disagreements between American and British strategists over the importance of Crossbow missions in comparison to other offensive missions against German industry. U.S. commanders argued that fast-moving fighter bombers were better suited to destroying the smaller targets and that it was a waste to use heavy bombers on anything other than the largest construction sites. The approach the Allies took resulted in a confused campaign against a relatively unknown threat. The campaign proceeded with uncertainty of its effectiveness, which illustrates the gravity of the situation as perceived by Allied war planners. A massive V-weapon attack on southern England could have wrecked Allied planning for D-Day.[29]

The aerial reconnaissance demanded by the War Cabinet on December 4, 1943, encompassed the whole of the Channel coast, 150 miles wide from southern England. The results startled Allied generals; in a matter of weeks, the number of ski sites had almost tripled, with 75 sites now identified. It was determined that bombing should begin as soon as possible. New tactical air forces were at that time being marshaled for

D-Day, and the U.S. Ninth Air Force was told to prepare for missions against the sites using medium bombers. They were joined by RAF aircraft and began the attacks on December 5, 1943, hampered by poor weather and visibility. However, it was almost impossible to discover whether the raids had been completely effective. Crossbow analysts were hard pressed to determine how many attacks would be required and the appropriate tonnage of bombs needed to knock out the V-1 sites. There was also a question about whether it was more effective to use high-flying heavy bombers carpet-bombing a large swath or small, fast-moving fighter bombers that could drop fewer munitions but with greater accuracy.[30]

After studying the initial results, Crossbow officials determined that medium bombers were not as effective as they had hoped. The fluctuating winter weather was not helping things either. Bombing missions over Germany were being postponed or delayed, sometimes for days on end; however, the shorter distance, along with better weather predictions, made it possible to attack targets in France. American generals, therefore, had no objections when on December 15, 1943, the British Chiefs of Staff asked for the use of the Eighth Air Force's heavy bombers in the Crossbow offensive. However, weather continued to be adverse, and it was not until the day before Christmas that the Eighth Air Force launched more than 1,300 aircraft during mission number 164, the first attacks on the ski sites. The sheer size of the mission attracted the attention of war reporters, and the secrecy of the operations against the V-sites was called into question that same day when the *New York Times* announced that U.S. and British flyers had hit the "Rocket Gun Coast." The editor commented, "The Germans have now created a diversion. They have at least won a breathing spell for themselves and temporarily diverted part of the Anglo-American air power. The threat alone has succeeded in lightening the weight of attack upon Germany."[31]

On December 22, U.S. Army Chief of Staff General George Marshall demanded more information about the potential threat posed by the German installations along the Channel coast. After he was briefed in Washington, Marshall questioned whether or not the British were completely forthcoming with intelligence concerning the Crossbow sites. He suggested an independent U.S. committee should evaluate the threat instead of relying solely on British information. A blunt memorandum was sent to British officials stating that U.S. Army Air Forces would offer no more assistance in Crossbow operations unless he was apprised of its importance. Supreme Allied Commander General Dwight D. Eisenhower and his staff had already been made aware of British concerns. Crossbow officials had asked Eisenhower's staff whether the German rocket campaign might adversely impinge on planning for Operation Overlord. Eventually, Supreme Allied Headquarters and the U.S. Crossbow committee came to about the same conclusions as did the British; the missile threat had to be taken seriously. If the Germans processed the ability to launch hundreds of V-1s each day, the insufficient number of antiaircraft guns could not have coped with such an onslaught. Missiles falling in southern England would adversely affect the Allied buildup for D-Day, especially if the missiles were directed at the port cities. The V-2

Flawless launch of an A-4 rocket from Test Stand VII sometime in 1943. Early test rockets were painted in an alternating black-and-white paint scheme, which made it easier to observe the longitudinal roll in flight. The waters of the Baltic can be seen in the background as the rocket rises. (*NARA*)

was said to be supersonic; there would be no countermeasure for it. The Allied generals figured it was not beyond reason that Hitler might use biological or chemical weapons against England to prevent the looming invasion of Europe.[32]

What the Allies could not have known was that Hitler's weapons of retaliation were still months away from deployment. Although accuracy of the missile was improving, with many of the late-summer test shots at Peenemünde falling within less than one kilometer of the target area, the V-2 troops still required live-fire training exercises, with range tables and computation data to be drawn up. Hitler had ordered all new test rockets to be launched in Poland. The only unit that had received any kind of V-2 operational training was Training and Experimental Battery 444, which consisted largely of VKN personnel from Peenemünde. The V-2 training program for new, qualified Army recruits had just begun. Enough equipment would be ready to outfit only one complete V-2 battery by December.

Leaving Peenemünde temporarily, Dornberger had established BzbV Heer headquarters near the Polish border at Schwedt on the Oder River. His staff was organized into three groups: a command group, a supply group, and an engineer group. Dornberger selected Lieutenant Colonel Thom as his chief of staff and ordered him to concentrate on problems related to the deployment of the missile in the field. In November 1943, with the multitude of complaints brought to Dornberger's attention about the innate deficiencies affecting the rockets and the general problems relating to the field employment of the V-2s, it was determined that a centralized office should be established to evaluate and coordinate important concerns. Dornberger placed Lieutenant Colonel Moser in charge and instructed him to compare all performance data and then pass it on to the appropriate section where remedial action would take place, with the most important matters being referred directly to Dornberger. This evaluation office proved to be the best way to communicate critical information throughout the rocket organization. The sources of mechanical and logistical break- downs were considered and corrective actions were taken. However, just as Dornberger was beginning this complicated process of training and organization, the command sit- uation changed again.[33]

The Commander in Chief of German forces in the west—Field Marshal Gerd von Rundstedt—had beseeched high command for combat control of the V-weapons. Since October 14, 1943, von Rundstedt had been responsible for defending the launching areas in the event of Allied invasion. In an attempt to have Dornberger relieved of his duties as field commander, Rundstedt proposed the creation of a spe- cial interservice corps under his authority for the tactical deployment of both the V-1 and V-2. On December 1, 1943, Hitler agreed with the plan and ordered the creation of a special Army Corps, Generalkommando LXV Armeekorps z.b.V. zur besonderen Verwendung (General Command 65 Army Corps for special purposes), combining the V-2 with the V-1 and long-range pressure guns to be used against England. The corps commander, Lieutenant General Erich Heinemann, an experienced artillery com- mander, was given overall command for active operations with both pilotless aircraft and long-range rockets.

After setting up LXV Corps headquarters in France, Heinemann made an immedi- ate assessment of the ongoing preparations for missile deployment. In light of the recent bombing campaign unleashed by the Allies in December, the massive hardened V-sites across the Channel coast seemed ridiculous to him. Heinemann believed the sites to be a tremendous waste of time and money, although he knew it would be point- less to openly criticize the extravagant building program authorized by the Führer. V- 2 site construction up to this point had followed the plan outlined by Dornberger and Hohmann in the summer of 1943; however, General Heinemann felt the installations were needlessly convoluted and easily identified from the air, which made them high- ly vulnerable to attack. The large V-2 site at Eperlecques lay in ruins, and during the raid on December 25, 1943, the Eighth Air Force had pounded the V-1 sites in Pas de Calais while the RAF bombed the V-sites at Abbeville. The overwhelming damage put an end to the immediate threat of V-1 attacks. Throughout the first months of 1944, the Allies continued to attack the sites from the air. Quietly, Heinemann ordered an overhaul of the entire deployment strategy, stressing more practical and easily camou- flaged constructions. Unable to repair the damage to existing sites, the Germans began building new firing positions, ones of a simpler, less distinctive style. Work on exist- ing sites would continue at a slow pace as a ruse.[34]

In January 1944, Dornberger set up HARKO 191 headquarters at the opulent estate Chateau de Maisons-Laffitte, just outside of Paris, near LXV Corps headquarters. However, because of recurrent failures plaguing the test firings, he spent most of his time either in Peenemünde with the scientists and engineers or in Poland with the trainees. Dornberger and Heinemann engaged in frequent arguments, more often than not regarding the overall readiness of the long-range missile.[35]

Heinemann believed Dornberger could better serve the program by concentrating on resolving the difficulties with the V-2 and also held the opinion that an officer with field experience should command the V-2 deployment. Even though Dornberger had spent the better part of his adult life in the pursuit of rocket development, this did nothing to sway arguments that he lacked experience in artillery operations in actual combat. Dornberger retained responsibility for overseeing technical development and troop training, but he was replaced as commander of the V-2 troops sometime in January. Dornberger's objections at his removal fell on deaf ears. Heinemann cast off the assertion that close coordination between the development team and troops in the field could only be maintained under Dornberger's direction. Major General Richard Metz, a novice to ballistic missiles, but one, who had commanded artillery with distinction on the eastern front, became the V-2 tactical commander. Metz was a no-nonsense commander. He was dismayed by the unconventional organization created by the young rocket enthusiasts. It was unlike anything he had experienced during his years of service in the German Army.[36]

As the year ended, the bilateral rocket assault on Great Britain seemed months, if not years, away. Production of the missile was just beginning at the Mittelwerk, the firing troops had just begun their training, the launching sites were being reorganized, and the command structure was anything but coherent. In September, there was a setback in V-1 production when the Fieseler factory at Kassel was bombed, causing extensive damage and delays in the program. The U.S. 15th Air Force attacked the Rax-Werke at Wiener Neustadt on November 2, 1943, causing extensive damage to the entire plant. Although Wernher von Braun had announced on September 9, 1943, that the development of the V-2 was complete, there remained many insufficiencies that made the rocket unfit for use; it could not be entrusted to units in the field, at least for the time being. Final blueprint drawings from Peenemünde had not been delivered to the Mittelwerk until early December. The underground production facility was in operation, but the quality and quantity of production was immature. The Allied air raids on Peenemünde and the Channel coast resulted in a complete reorganization of rocket production and deployment strategy, one that depleted valuable time and resources. Those involved had to admit to themselves there was no possibility of a rocket offensive against Britain until the following summer, if then.[37]

If there was one person who was quite pleased with the state of affairs surrounding the long-range rocket, it must have been Reichsführer SS Heinrich Himmler. On September 29, 1943, he flew to the SS proving grounds in Poland to make the necessary arrangements, in person, for the transfer of rocket testing and training from

Peenemünde. In less than six months the SS had managed to snare and then entangle itself in the rocket program. Himmler now controlled rocket production and soon would wrestle even more power from the Army.

4

Preparations for Vengeance

Following the bombing of Peenemünde in August of 1943, missile test firing was moved from Peenemünde to the southeast corner of occupied Poland. Established in 1940, an SS training and proving ground, code-named Truppenübungsplatz Heidelager (Troop Training Ground Heidelager), was situated at Pustków-Blizna some 90 miles northeast of Krakow, exceeding the range of Allied bombers. Blizna was a very small village in the middle of nowhere. All of the Polish inhabitants had been evicted when the SS established the 12-square-mile training camp. A new railway line had been constructed to the village: the Mielec-Debica-Tarnow rail line, connecting to the Krakow-Lemberg line, which had been extended. A new concrete road connected the camp to the nearest main highway. Approximately 12,000 SS soldiers were living and training at Heidelager at any given time during the war; however, the portion of the camp given to rocket testing was newly constructed in the center of the main SS compound. The secure launching area measured about one square kilometer and was well hidden in the middle of the forest. Even the SS soldiers guarding the surrounding camps could not enter the restricted rocket range. At Heidelager there also existed a camp for Russian POWs and a concentration camp that housed several thousand Jews. Artilleriezielfeld Blizna (Artillery Range Blizna) was maintained by the SS, but it was 400 men from the army that arrived in October to construct the various buildings, stands, sheds, and roads needed to support rocket-testing operations.[1]

The only buildings that stood in the area were a single white-painted wooden house with a small stable and, a bit farther, a stone schoolhouse. To these were added a long assembly building near the railhead along with living quarters, storage sheds, and workshops of various variety, all surrounded by a double barbed-wire fence. Officers were housed in a nearby parked train with relatively plush amenities compared to the enlisted men's barracks. Shipments of machinery and equipment arrived soon afterward. A wooden proofing tower, similar to the towers used at Peenemünde, was constructed approximately 100 meters from the railhead and assembly building.

Measuring about 16 square feet at its base, the tower rose to a height of 50 feet with servicing platforms approximately every eight feet. One of the first concrete firing platforms was constructed about 300 feet east of the tower. All locations were connected by a Decauville light railway track. Secrecy was, again, a high priority. Fake houses were constructed around the camp, complete with dummy farm animals and laundry hanging out to dry, all meant to conceal the true purpose of the activities at Blizna.

In September 1943, Peenemünde's Colonel Gerhard Stegmaier organized the long-range rocket unit activities and opened a training school for the new V-2 troops near Köslin. The training consisted of two courses: Schedule A for officers and those with engineering experience and Schedule B for ordinary soldiers without the need for technical knowledge. Stegmaier also organized the Artillerie Ersatz Abteilung 271 (Training and Replacement Detachment 271) at Schneidemuehl, which operated as the field instructors for the designated battalions and also would provide qualified replacements for any potential casualties in the field. The training of the technical units for the operational field stores was carried out under the direction of Lieutenant Colonel Basse at Peenemünde.

Heavily guarded trains began arriving at Blizna, the carriages camouflaged with large tarps and a new type of tanker car in tow. The first battery to complete their training at Köslin was the Training and Experimental Battery 444. Under the command of Major Wolfgang Weber, the men of Battery 444 arrived at Heidelager and readied six mobile firing locations. In the first few weeks, troops practiced setting up the firing platform, bringing around the Meillerwagen trailer, erecting the rocket, and moving up the fueling vehicles. Battery 444, which was only one battery (three firing platoons), was later joined at Heidelager by Artillerie Abteilung 836 (Artillery Battalion 836) and Artillerie Abteilung 485; each was a battalion having a headquarters battery, a technical battery, and eventually three firing batteries (nine firing platoons).[2]

The first V-2 launch organized at Heidelager by Colonel Stegmaier and Major Weber was a failure. The temperatures that morning—November 5, 1943—were well below freezing as the firing crew readied the rocket in the center of a sandy road. Owing to inexperience or carelessness, a Battery 444 soldier had not lowered the blast deflector plate on the firing table completely to the frozen ground. At preliminary ignition, the hot gases quickly thawed the ground, causing one of the table legs to sink into the wet soil. The rocket canted at liftoff, went out of control, and crashed several miles away. It was easy to understand the cause of the failure. However, LXV Corps commander General Heinemann was present and witnessed the crash. Immediately, he incorrectly concluded that the rockets could only be launched from concrete pads. Even after it was proven that the rockets could be launched from almost any firm surface, orders were issued to build hundreds of concrete pads in the planned deployment area of northern France. Much time and effort was wasted in the construction of these concrete emplacements that were never used.[3]

In December 1943, the first battery of Artillery Battalion 836 moved from Grossborn near Stettin to Peenemünde. The trainees were assigned duties in Test Stand

VII, and after a series of introductory lectures, their real training commenced. Among the courses given were general engineering; surveying, or Vermesser; power plants, or Triebwerk; and a basic electrical course. Instruction, along with practical application, was conducted in a large workshop just southwest of P-7. Around mid-January 1944, the whole of Battalion 836 moved from Peenemünde to Heidelager for further instruction and firing practice. Upon arrival, they found that crews from the Training and Experimental Battery 444 were already there; however, Battery 444 was not engaged in operational training of its own but

V-2 is pulled from storage shed on Decauville light railway track at the Heidelager SS training ground near the town of Blizna in occupied Poland. After the RAF raid on Peenemünde on August 17, 1943, it was decided that the training and testing should be done in southeast Poland out of the range of the Allied bombers. In July of 1944, the advance of Soviet troops forced the base at Blizna to be evacuated, and launch activities were moved to the Tuchola Forest. (*NASA*)

rather was conducting experimental shoots in conjunction with those still being done at Peenemünde. In addition, Battery 444 was responsible for providing the instructors and equipment for the training of the new batteries, which included the batteries of the 836 as well as its sister unit, Artillery Battalion 485. At Heidelager, the first battery of Battalion 836 fired four rockets. Of these, one fell on the launching position after having climbed a few meters, one was completely successful, and the other two deviated from planned course, necessitating the activation of emergency fuel cutoff. The training at Heidelager lasted about six weeks, and then the batteries returned to Grossborn in Poland where they were equipped with the numerous specialized vehicles required for independent V-2 operations.[4]

General Dornberger was looked upon favorably by the rocket crews at Heidelager. He was a frequent visitor to the launching sites, not the least bit timid, and would don a pair of overalls to jump into the mix during firing preparations. The 48-year-old Dornberger took seriously his duties for forming and training logistical field employment methods. The troops, fresh from the V-2 school at Köslin, needed realistic training in all weather conditions. The training was not limited to firing procedures, it also involved fuel trains, technical batteries, repairs in the field, security, and vehicle maintenance. Dornberger was pleasantly surprised to discover many helpful suggestions coming from the soldiers, many of whom already possessed a limited background as craftsmen and engineers in civilian life. The A-4 Fibel (V-2 field handbook) was soon introduced at Heidelager. It was used to instruct the rocket troops about the mechanics of the V-2 and their individual duties.[5]

The testing of integrating accelerometers for V-2 engine fuel cutoff took place at Heidelager. Its introduction was a major breakthrough because it eliminated the need for a ground connection in any direction to achieve correct timing for Brennschluss (shutting down the engine). This component was mounted in the rocket with the ability to "sense" acceleration. As soon as a preset velocity was attained during flight, the device would switch off the flow of fuel to the engine. The first devices tested in flight did not deliver consistent results; however, the performance was relatively similar to the ground radio cutoff system. Always somewhat suspicious that the ground-based radio equipment might be susceptible to Allied jamming techniques, Dornberger accepted the performance of the accelerometers and was happy to discard the expensive ground equipment previously needed for timed fuel cutoff. The firing crews at Blizna continued to hone their skills of moving, fueling, and firing a V-2 from a remote location in various types of terrain. Considering the doubts cast by some of the technicians at Peenemünde, Dornberger was delighted to see the soldiers quickly became proficient in all operations.[6]

In the spring of 1944, instead of keeping prisoner workers in tunnels, Dora was changed into a typical Nazi slave labor camp, with 58 barracks surrounded by barbed wire being set up about a quarter mile west of the south entrance to Tunnel B of the Mittelwerk. Total camp construction was not completed until October, 1944. The SS had set up a large network of smaller concentration camps to supplement camp Dora; these consisted of subcomponent depots and workshops related to rocket production all near the town of Nordhausen. Project B12, alias Kaolin, was a Kohnstein tunneling project west of the Mittelwerk tunnels, intended to create an even larger underground factory (160,000 square meters versus 125,000 for Mittelwerk). The plan foresaw Tunnels C-1, C-2, D-1, and D-2, parallel to Mittelwerk's A and B, being used for an aircraft factory. Project B11, alias Zinnstein, was situated on the eastern side of Mittelwerk. It had two galleries connecting with Mittelwerk at the level of Halls 17 and 43. A checkerboard pattern of underground chambers was planned, with 80,000 square meters of usable space. Three factories were to occupy this space: one for synthetic fuel production, one for liquid oxygen production, and another for aircraft manufacture. A final project called B3, or Anhydrite, in the area west of Himmelberg, included a checkerboard pattern of 28 tunnels connected by perpendicular halls. Its 130,000 square meters were to house another aircraft factory. It is suggested this plant was intended to handle production of the R4/M Orkan, an air-to-air, fin-stabilized antiaircraft missile; the Taifun, a surface-to-air antiaircraft barrage rocket; and the X4 air-to-air and X7 air-to-ground wire-guided missiles. Electrical power was produced by hydrostations at Bleicheröde and Sondershausen.

Since the camp at Dora could not hold all the prisoner labor needed for all these new projects, General Kammler decided in March 1944 to create two new concentration camps: Harzungen, on the east side of the project area, and Ellrich, next to the train station in the village west of the Mittelwerk. These two camps had a higher death rate among their inmates than did Dora. Since October of 1943, the number of pris-

V-2 wrapped in camouflage tarpaulin on Vidalwagen road transport trailer. The rocket is being pulled by the Fiat SPA TM-40 artillery tractor. The Vidalwagen was used to move the rockets from the railhead to the technical troop field store and then again to the firing troop transfer point where the missiles were hoisted by crane onto the Meillerwagen. It was lightweight but very strong. The 46-foot-long tubular design could easily support the 4,485-pound weight of the empty V-2. It was designed and manufactured by the Vidal & Sohn Company. (*NARA*)

oner workers at Dora had risen from 7,000 to over 12,000 in January of 1944, with an additional 15,000 slave laborers in the Dora subcamps as well. Chillingly accurate SS records show that during a period of seven months, from the end of August 1943 to the beginning of April 1944, more than 17,000 detainees were shipped to Dora. However, the number of prisoners counted by the SS at the beginning of April was under 12,000. Thousands of persons had already perished in the tunnels and were incinerated in the ovens at Buchenwald. Others too sick to continue working were transported to other camps, presumably to die there.[7]

The expanded Mittelwerk tunnel system consisted of two parallel main tunnels— A and B—each roughly 6,200 feet long, bent in a shallow "S" curve, and connected at various points by a regular series of cross tunnels, like the rungs of a ladder. The cross tunnels, called Kammer (halls), were about 600 feet from the outside wall of Tunnel A to the outside wall of Tunnel B and were numbered from 1 to 46, beginning at the north side of the mountain. Tunnels A and B had a height of 21 to 23 feet and a width of 29 to 36 feet. The cross section halls were somewhat smaller; however, the total estimated expanse was over one million square feet. A small production line for the V-1 flying bomb occupied the southern tip of Tunnel A, the first four halls, and the entryway (Halls 43 through 46), which had been used as the prisoner's sleeping quarters before the construction of the Dora camp barracks outside the tunnels. V-2 assembly, known as Mittelwerk I, occupied Halls 21 through 42, while the northernmost end of the complex, Nordwerk, was designated for Junkers jet engine production. The office of the director, Alwin Sawatzki, was situated in the middle of the factory. Each of the main tunnels could accommodate the entry of full-scale railway transport trains.

Tunnel A served as the entry portal for materials heading into the factory. Tunnel B served as the main assembly line, beginning at Hall 21 and moving south toward Hall 42, until the completed rockets left the mountain via the Tunnel B exit. There were two parallel assembly tracks in Tunnel B, which carried rockets on Decauville track bogies from north to south. The assembly line finally arrived at Hall 41, a gargantuan chasm more than 50 feet high, which had been excavated far below the floor level. Here a massive spanning crane allowed the rockets to be erected in the vertical position for final inspection. One whole side of Hall 41 contained a series of multilevel scaffolds for vertical inspection of the rockets, simply because final fluid and gyroscopic tests could only be carried out on a vertical rocket.

The Mittelwerk assembly lines included not only slave laborers but also German workers and supervisors. Trained engineers ran the workshops, while German soldiers acted as inspectors. At any given time, the ratio of slave laborers to German workers was approximately 2 to 1. The prisoners were divided into two work groups: transport/construction and specialists. The former did the often back-breaking work of manually transporting much of the material that entered or left the tunnels, while the latter did other more skilled assembly and testing work. They were divided into day and night shifts, each working 12 hours straight with no breaks. Kammler's SS henchmen were integrated all over the plant, secretly monitoring all aspects of production.

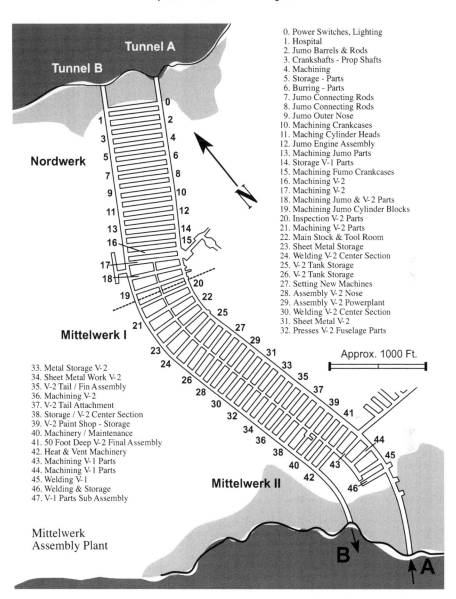

0. Power Switches, Lighting
1. Hospital
2. Jumo Barrels & Rods
3. Crankshafts - Prop Shafts
4. Machining
5. Storage - Parts
6. Burring - Parts
7. Jumo Connecting Rods
8. Jumo Connecting Rods
9. Jumo Outer Nose
10. Machining Crankcases
11. Maching Cylinder Heads
12. Jumo Engine Assembly
13. Machining Jumo Parts
14. Storage V-1 Parts
15. Machining Fumo Crankcases
16. Machining V-2
17. Machining V-2
18. Machining Jumo & V-2 Parts
19. Machining Jumo Cylinder Blocks
20. Inspection V-2 Parts
21. Machining V-2 Parts
22. Main Stock & Tool Room
23. Sheet Metal Storage
24. Welding V-2 Center Section
25. V-2 Tank Storage
26. V-2 Tank Storage
27. Setting New Machines
28. Assembly V-2 Nose
29. Assembly V-2 Powerplant
30. Welding V-2 Center Section
31. Sheet Metal V-2
32. Presses V-2 Fuselage Parts

Approx. 1000 Ft.

33. Metal Storage V-2
34. Sheet Metal Work V-2
35. V-2 Tail / Fin Assembly
36. Machining V-2
37. V-2 Tail Attachment
38. Storage / V-2 Center Section
39. V-2 Paint Shop - Storage
40. Machinery / Maintenance
41. 50 Foot Deep V-2 Final Assembly
42. Heat & Vent Machinery
43. Machining V-1 Parts
44. Machining V-1 Parts
45. Welding V-1
46. Welding & Storage
47. V-1 Parts Sub Assembly

Mittelwerk
Assembly Plant

Each prisoner work group was under the supervision of a prisoner leader, or Kapo, who was a prisoner appointed by the SS command. Many of the transport prisoners were required to move heavy material from the outdoor storage facilities into the Mittelwerk assembly halls. If weak prisoners fell down, the SS guards and Kapo were there to kick and beat them with truncheons until they could lift their burdens and continue once again. Much of the work was done in the dead of winter, the workers sloshing through snow, ice, or mud, with nothing on their feet other than mitts or clogs.[8]

The groups of prisoners were generally organized by nationality—the better-educated French workers, who where mostly civilians arrested by the Gestapo for political reasons, ended up at jobs like electrical assembly and testing. Many of the French prisoners who possessed engineering degrees or had specific technical training were hand picked by Peenemünde officials for specific jobs at the underground complex. The transport and construction groups, on the other hand, tended to be made up of ordinary Russian and Ukrainian prisoners. The specialists were organized into various Kommando (work groups) assigned to workshops, assembly points, or subcontractors near the complex. Some 20 German companies were involved in the Mittelwerk construction and missile assembly process. They included such well-known names as Siemens, AEG, Telefunken, Rheinmetall, Ruhrstaal, BMW, Junkers, Heinkel, Walther, Askania AG, and DEMAG.[9]

Even though the tunnels were vast, space in Mittelwerk was limited; only the most delicate parts such as gyroscopes and electronics were stocked completely underground. Bulky objects like motors, fuel tanks, and half-fuselage sections were stored aboveground in smaller neighboring depots. One of these was situated at the north end of Tunnel A in the town of Niedersachswerfen and was crewed by a Czech Kommando. The second depot was at Rossla, about 21 kilometers east of Nordhausen, in the old Kalkofen factory. The third depot was located at Kelbra. In late 1944, a small Kommando would also be located in an old potash mine at Kleinbodungen, near Bleicherode. This crew worked to refurbish damaged V-2s that had been returned as a result of misfirings, damage during transport, or other factors that made them unusable by the launching batteries in the field.

The Mittelwerk's first 20 units were sent by rail to Peenemünde for evaluation and testing at Prüfstand VII. After production was underway, the number of man-hours required to produce one rocket was around 15,000. After the plant had been in operation for several months, the number of man-hours to make one rocket should have been reduced by half, to about 8,000 hours. All the heavy machinery had been installed and workers were busy on the lines when Wernher von Braun visited the Mittelwerk for the first time on January 25, 1944. The vast expanse of the workspace—over 97,400 square meters—certainly must have impressed him, but what about the appalling living conditions of the slave laborers? Prisoners were dropping at an astonishing pace— over 50 each day on average.[10]

Until 1944, rockets launched from Peenemünde had been from the Versuchs series—a total of 46, all of which had been manufactured by Peenemünde technicians and engineers. In January of 1944, Peenemünde and Heidelager began receiving the first group of rockets manufactured at the Mittelwerk. Shipped by rail, the rockets arrived on flatcars. Since each rocket was longer than one railcar, it required the use of three railcars to transport two rockets. Placed nose to nose, the rockets were surrounded by a huge tarpaulin supported by a framework. The canvas cover not only shielded the rockets from view but also protected them from the elements. The warheads and accessories would later be carried on the middle flatcar between each pair of rockets. Upon examination, Peenemünde experts found numerous problems with the initial batch of rockets. The first series production rockets carried the serial numbers 17001,

17002, 17003, etc. Number 17003 was the best of the lot, but many obvious mistakes would need to be corrected by Peenemünde technicians. Problems included cracked seals, misfitting parts, broken struts, leaks, and wrong connections. Finally, after many delays, 17003 was prepared for launch from Test Stand VII on January 27, 1944. Searchlights directed their beams to the rocket as it lifted off late in the evening. After rising several feet, the engine shut down, the exhaust flame vanished, and 17003 fell back onto the firing point, rocking the complex with a massive explosion. One of the first units manufactured at the Mittelwerk had failed, and in doing so, it had damaged the launching pad at Test Stand VII. Later, as production quality improved, fewer missiles had to be shipped to Peenemünde for examination.[11]

If the experts at Peenemünde were having difficulties when launching series production rockets, even after technicians had made extensive corrections, one can imagine the frustration level at Heidelager. The operational training of the launching crews was going well; however, all were surprised and disappointed in the performance of the missiles. In succession, failure after failure baffled the launch crews. Very few of the rockets were completely successful. Rockets were falling back on the launching point after rising only 50 to 100 feet. Ground support equipment, which was in short supply, was being destroyed. Cables were severed, firing tables were mangled, and in one of the worst cases, a falling rocket killed 23 soldiers of Battery 444. Other rockets would rise to almost a mile in height before mysteriously exploding, giving no clue as to the cause. When a round managed to make it successfully to the target area, a loud double bang was heard, followed by the warhead impact and then a shower of metal fragments in the target area. These particular shots were breaking up before reaching the ground.

Some of the Peenemünde personnel were blaming the inexperienced firing crews, but even after integrating Peenemünde technicians within the crews, the problems continued. There was a joke among the rocket troops: "Let's make the target area more dangerous than the launching sites." The crews had serious doubts whether the rockets would ever perform as prescribed. Peenemünde personnel were sent to the Mittelwerk to inspect the manufacturing procedures. The brass was getting impatient, and modification after modification was quickly introduced. There were meetings followed by more meetings where every possibility was discussed, but nothing gave a clue to the source of the failures.

After exhaustive examination of existing clues and possible remedies, the problem of explosion occurring during liftoff was virtually eliminated. It was found that the fittings at each end of the curved fuel pipes were loosening from the jolt at ignition and during the heavy vibrations of thrust at liftoff. The high pressure in the pipes from the turbo pump caused a fine mist of fuel to spray from the loose fittings into the tail section, which was ignited by the exhaust flame. Improved pipes and extra locking devices for the fittings solved this problem. It was also discovered that the heavy vibrations throughout the rocket body during liftoff were shaking loose a set of relay contacts that distributed power from the onboard batteries to the subsections of the rocket, resulting in premature engine cutoff. After the correction of these problems, most of the rockets lifted off without failure; however, there was still the problem of rockets disintegrating upon reentering the dense layers of the atmosphere. The airburst problem was not unique to rockets manufactured at the Mittelwerk. Rockets built at Peenemünde were meeting the same demise. Only 30 percent of the missiles came down intact or impacted anywhere close to the aiming point. It was a big mystery. Dornberger could only guess that it might be due to structural weakness caused by external heating or an explosion from excess fuels in the tanks. The few remaining fragments were collected from the ground to be examined for clues.[12]

By early 1944, the U.S. Crossbow committee came to essentially the same conclusion as the British: hit the V-weapons sites before they could become operational. However, there was much debate among the committee members over whether diverting the four-engine heavy bombers away from their primary mission of smashing Germany's heartland was the best strategy. Almost half of the missions flown by the Eighth Air Force in January and February of 1944 were directed against Crossbow targets, while in staging areas all over England there were thousands of fighter bombers and medium bombers waiting for the beginning of Operation Overlord. Some committee members argued that these idle aircraft could be utilized more effectively against the V-1 ski sites and the V-2 bunkers and at the same time, free up the heavy bombers for a continued assault on German industry. The committee asked the USAAF to conduct tests to determine the most effective method of hitting the V-sites in France.[13]

In response, on January 25, 1944, the commander of the U.S. Army Air Forces, Henry H. "Hap" Arnold, ordered the bombing of Florida! General Arnold wanted tests to be conducted using virtually every type of aircraft in the Allied air arsenal—fighters, medium bombers, and heavy bombers—all of them flying against mock ski sites quickly constructed at the USAAF Proving Ground Command at Eglin Field in the Florida panhandle. The attacks would be carried out in war-game scenarios against real antiaircraft units, and the mock sites would be heavily camouflaged.

A wave of Army purchasing agents from Eglin fanned out across the state to buy all available concrete, cement, and steel from wherever it could be found. The project was soon the largest building project on the eastern seaboard of the United States. Workers labored around the clock to complete the work. Twelve days after work began, the project was complete, and by the first week of February, the bombs were falling. The mock V-buildings were being smashed as soon as the concrete hardened. Air Corps experts determined the destructive force of various bomb types, the ease and accuracy of the aircraft delivering them, and the success or risk of various attack profiles. The results showed that quick attacks coming in fast at low altitudes provided the best results. This method gave the ground defenses very little warning, and although the amount of bombs placed on the target was minimal, the accuracy was substantially improved over high-level attacks.

On February 19, 1944, General Arnold attended an actual demonstration at Eglin Field and was thoroughly convinced. Eglin officials prepared an official report detailing the conclusive findings of the trials, which proved that low-level attacks, using fast fighter aircraft such as the American P-38 or the RAF Mosquito, were just as effective as the four-engine heavy bombers. The report went on to tout the use of fighter bombers, highlighting the substantial cost savings and reduced risk of pilot loss. The report was delivered to Supreme Headquarters Allied Expeditionary Force (SHAEF) on March 1, 1944. The circulation of the test results caused rancorous debate within the Allied command. The British refused to accept the results and continued to favor high-altitude bombers. The RAF believed smaller targets would be more easily and safely hit by high-altitude area bombing. General Arnold was infuriated at what he alleged to be British unreasonableness and an apparent weakness for hypothetical theorizing rather than believing hard evidence. Despite rising resentment among American air chiefs in Washington and Britain, Supreme Allied Commander Eisenhower acquiesced to the desires of the War Cabinet, which continued to insist on the British approach. Consequently, the Allies dropped 4,200 tons of bombs during the more than 2,500 Crossbow-related sorties in March.[14]

Following the incessant bombing of the ski sites in France, LXV Corps commander General Heinemann ordered a new type of modular launching ramp for the flying bombs. Construction of the new emplacements was very limited and only required the pouring of 16 concrete support blocks and a trolley platform. A new prefabricated steel launching ramp could be erected in sections and mounted on the supports in a matter of days. These new sites were easy to build, many in number, and easily camouflaged. LXV Corps completely abandoned the old sites and supply depots, which were built prior to the formation of LXV Corps and recognized by the Allies as targets. The target date for beginning the flying bomb offensive had been postponed several times and was now slated for the end of April. A sense of urgency permeated the German High Command because the Allied invasion was anticipated at any moment.[15]

On April 18, 1944, the British War Cabinet urged General Eisenhower to step up attacks against all suspected V-sites. The British finally acknowledged that the campaign up to that point had shown poor results, but this was only because, in their opinion, insufficient resources had been allocated to the campaign. Now they demanded more. Eisenhower deliberated and, the next day, granted Crossbow missions priority over all other air operations. By the end of April, Crossbow had increased in volume; a total of 7,500 tons of bombs had been dropped on the V-sites during over 4,000 sorties. Even though they didn't like it, American air commanders were forced to defer temporarily to Crossbow. There was a growing concern that the Allies might not have enough aircraft to support Operation Overlord, much less the daylight campaign against Germany's heartland.

General Arnold was unwilling to endorse Crossbow over other strategic bombing offensives. He set out to prove the results from the trials at Eglin Field Proving Grounds by ordering low-level attacks on four ski sites. Four P-47 fighter bombers, each armed with two 1,000-pound bombs, made pinpoint raids, scoring Category A damage despite heavy antiaircraft fire. As D-Day drew closer, U.S. officials pressed for the release of the heavy bombers for other strategic operations. The Americans had pummeled ski sites with Category A damage 107 times, and most of the sites were ruined. The statistics coming in supported the U.S. contention that the use of heavy and medium bombers was wasteful. It was hard for the British to dispute these results, especially when in all of the RAF-flown Crossbow missions—they had never scored Category A damage.

Meanwhile, Reichsführer SS Himmler, with V-2 production firmly in his grasp, soon made an attempt to gain control over the Army's research center at Peenemünde. One evening in February 1944 (the exact day is uncertain), he summoned Wernher von Braun, who was an SS officer, to SS headquarters near Rastenburg. The next morning, von Braun climbed aboard his state-owned Me-108 airplane and made the journey to East Prussia. Upon arrival, he was chauffeured to Himmler's headquarters, which was located in a camouflaged train at Hochwald. Even though Himmler had ordered the young technocrat's appearance, the Reichsführer SS began their conference as if it were some pleasant tête-à-tête between old acquaintances.[16]

In his memoirs, von Braun wrote that after some peculiar niceties, Himmler came to the point. "I hope you realize that your long-range rocket has ceased to be a toy for your amusement and that the whole German people eagerly await the weapon of retaliation. As for you, I can imagine that you have been immensely handicapped by the hidebound Army bureaucracy and red tape. Join my staff here at the SS, and we'll really get this project going. Surely you must know that no one has greater access to the Führer, and I can promise you much more effective support than you will ever get with those Army generals."

Von Braun responded quickly (maybe, in retrospect, a bit too hastily), "Reichsführer, I could not ask for a better boss than General Dornberger, and our delays are due to technical troubles, not Army red tape. We are almost there, and any change to the program at this juncture would be downright disastrous."

When von Braun had concluded, Himmler changed the subject for a few seconds and then dismissed the youthful professor with a peevish smile. Himmler was not easily rebuffed. Without qualified scientists within their own ranks, Himmler believed, for the SS to acquire complete control of the rocket program, he must also control the research arm of the organization. Hitler was growing impatient with the delays involving technical problems with the V-2, and the Reichsführer SS believed the time had come for him to make a move.

On March 5, 1944, von Braun was relaxing at a party in the town of Zinnowitz, near Peenemünde, after a hard week at the research center. Several of his associates from research center were in attendance, among them Klaus Riedel. Von Braun had a couple of drinks, relaxed, and played several classical pieces on the piano. Von Braun's account, which has been repeated many times, relates that at some point in the festivities, the young rocket scientists began talking openly about their "consuming passion" for space travel but were unaware that their conversation was being carefully monitored by Gestapo agents at the party. A few weeks later, von Braun was awakened in the early morning hours by a loud pounding on the door of his rented room in Koserow, 19 kilometers south of Peenemünde. When he opened the door, he was immediately confronted by three Gestapo agents and was driven to a prison in Stettin. In reality, many of Peenemünde's department heads had been under surveillance by Gestapo agents long before March 1944. The SD (SS intelligence service) had begun compiling dossiers on von Braun and his colleagues as far back as October 17, 1942. Also brought in were Klaus Riedel, von Braun's brother Magnus, and a few other Peenemünde associates. They had been accused of sabotaging rocket development by focusing their efforts on spaceflight rather than weapons for the Führer and, according to the personal diary of General Jodl, of possibly constituting a "refined Communist cell."[17]

Von Braun said later, "Finally, a court of the SS charged me with statements to the effect of saying the rocket was not intended as a weapon of war, that I had space travel in mind when it was developed, and that I regretted its imminent use as an operational weapon. That wasn't so bad, since many of the Peenemünde personnel felt this way; however, when they charged me with intentions of maintaining an aircraft at the ready to fly to me to England with important rocketry papers, this would have been difficult for me to disprove."[18]

The seriousness of the situation was far graver than von Braun knew at the time. Dornberger, who was on his way to Bertesgarten, found out about the incident late in the evening when General Buhle told him what had happened. Upon hearing the news, Dornberger was astonished. The following morning, Dornberger reported to the office of Field Marshal Keitel, who informed him that the charges were very serious and that the men in custody could be executed. Dornberger said he would personally vouch for von Braun and Riedel regardless of the charges against them. He demanded that the prisoners be transferred from the Gestapo to Army control, because civilian employees of the Army were under the jurisdiction of military courts. Keitel was extremely afraid of crossing Himmler and offered no help. Dornberger then attempted to get an appointment with Himmler himself only to be turned down. Unrelentingly, Dornberger then drove to the SS headquarters in Berlin. There, he had to deal with SS General Heinrich Müller, chief of the Gestapo, who was filling in for an absent SS General Kaltenbrunner.

Müller began splitting hairs at once. He pointed out that Dornberger's men had not been "arrested," they had been taken into "protective custody." Dornberger fumed that as far as he was concerned, there was no difference between the two. Upon hearing such statements, Müller informed Dornberger that he had quite a dossier on him as well and many other Peenemünde personnel. Dornberger deciphered from Müller's rambling comments that SD agents were most probably on his own personal staff! He bluffed Müller by saying that if the SS thought the charges would stick, he should be immediately arrested. It was a dare that Müller couldn't accept, with so much fanfare being raised in the Nazi Party about the miracle weapons and how they would bring ultimate victory to the Reich. After several days of tedious negotiations between the Army and the SS, Dornberger secured the release of von Braun and the others. But unbeknownst to Dornberger, Albert Speer had spoken to Hitler and convinced the Führer that von

Braun was indispensable; there would never be a V-2 if he remained under arrest. If Hitler had truly believed that the project could have seen its way to fruition without the help von Braun, he probably would have let Himmler dispose of the newly appointed professor. Von Braun's arrest, which later was seen by many as proof of his resistance to Nazi beliefs, was, in reality, not related to any concerns about his true intentions for the rocket or whether or not he held any Nazi beliefs. In fact, Himmler organized the whole incident. Nobody snubs the Reichsführer SS, and if von Braun would not conspire with Himmler to make the SS the absolute authority over the rocket program, then he should be eliminated.[19]

In May of 1944, von Braun met with Dornberger at Heidelager. Pleased to be back on the job, he had come to help solve the mystery of the mid-air breakups. They flew in a light aircraft 164 miles northeast to the test target location near the Polish town of Sarnaki to examine the fragments of charred wreckage collected after the rocket airbursts. A few weeks earlier, about 40 German soldiers of the observation team had been stationed at Sarnaki, south of the Bug River in a local schoolhouse. There, the chief of the observation squad led them to a building in the village that housed the remnants of the missiles fired at the target area. Inside they found smashed tail sections, combustion chambers, piles of twisted feed pipes, mangled aluminum pieces from shredded fuel tanks, gyroscopes, electrical wiring, and crumpled sections of the outer metal skin and frame. It seemed almost every section the complete assembly was represented in some fashion. Dornberger examined a small piece of the control compartment outer skin, which was fashioned to a piece of wood framing. The frictional heating of over 1,200 degrees Fahrenheit during the rocket's return through the atmosphere had charred the piece of metal, but seemingly left the wood unaffected. How did this tiny piece of charred metal relate to the monumental difficulties they were having?[20]

It was near this small town that a V-2 almost ended the lives of Dornberger and von Braun. On a warm day, late in the evening, the two men had been milling about in the target area near an observation tower. Von Braun happened to look up at the time indicator on the tower to notice a V-2 launched from Blizna some minutes earlier was about to arrive in the area. He turned and peered up at the sky in the direction from which the rocket was converging. He was shocked to see a thin, faint contrail falling in their direction. He and Dornberger barely had time to fall face first onto the ground before a terrific explosion ripped the earth only a scant 90 meters away. They both were hurled into the air, von Braun coming to rest in a nearby ditch. It was a miracle that neither man had been injured or killed by the impact or the exploding warhead.[21]

Undaunted by their brush with death, Dornberger and von Braun ventured to the target area again several days later. From a small observation trench, the men anticipated the arrival of the next round from Blizna. Up to this point, some 60 rockets had come down near Sarnaki, most of them in thousands of pieces. As human targets, they hoped to catch a glimpse of an incoming rocket. As remote as the possibility might have been, the experience of a few days earlier had revealed it was possible to witness an incoming rocket, if only for a few seconds. Maybe they could observe something that would yield a clue as to the cause of the breakups. A radio communication from Blizna announced the firing of a rocket. Dornberger clicked his stopwatch. It should take the rocket a little over five minutes to travel the distance from Blizna to the observation post. Dornberger raised his binoculars at the five minute mark and searched the heavens in the direction of the incoming missile. He was surprised to see what he thought was a fast moving streak surrounded by a faint burst of steam. There was some type of separation in the vehicle and almost immediately thereafter, an ear-splitting double boom as the warhead slammed into the earth about a mile ahead of them. As the dust cloud from the impact dissipated, large chunks of rocket debris slowly fell to the ground. Dornberger suspected he had just witnessed an airburst. They drove to the point of impact and examined the wreckage, but it gave no additional clues.

The men flew back to Heidelager, where exhaustive discussions were held with General Erwin Rossmann and senior Peenemünde staff as to the cause of the breakups. Rossmann was new to Peenemünde, installed by Army Ordnance to head the liquid-fuel rocket section, but he offered some helpful insights. Rossmann proposed that the tank sections of a half dozen rockets be insulated with glass wool to prevent any potential heat transfer from weakening the outer frame. This was a potentially easy solution to a perplexing problem. If it worked, the fix required almost no modification to the rocket. The group also decided that several more rockets should be tested, each with only a modest amount of alcohol so that the tank would run dry in flight and be empty upon reentering of the atmosphere. If there were no more explosions or breakups, this could isolate the tank as the root source. These rockets would be launched without a ballistic trajectory, going straight up, from the little island of Greifswalder Oie off Peenemünde. Due to the rotation of the earth, they should fall only a few kilometers offshore and be easily photographed upon reentry. Newly installed antennas in fin 1 and fin 3 would send signals from an onboard transmitter using 24 channels to monitor multiple suspected points during flight.[22]

The rockets flown with minimal alcohol gave no definitive conclusion. However, the six rockets with the tank sections wrapped in fiberglass insulation all came down intact. Whether the insulation dissipated the heat transfer on the mid-section or stiffened the frame enough to strengthen the center of gravity, the fact that all rockets with the tank insulation performed correctly without a breakup was all the engineers needed to know. Future tests would show the fiberglass did not solve the problem entirely, but the improvement was enough to go forward with deployment of the missile. In another effort to surmount the mid-air breakups, workers at the Mittelwerk added steel reinforcing sleeves that were riveted around the fuel tanks. It was soon discovered that most of these rockets reentered the atmosphere without incident and impacted the target area.[23]

Peenemünde also continued the evaluation of series production rockets. On March 17, 1944, one of the Mittelwerk examples was launched from P-7 late in the morning. Ignition and the initial ascent looked good. The rocket rose straight up, began the programmed ballistic tilt, but after reaching about 1,000 feet, the engine shut down. The rocket hung in the air; the roar went silent. It turned over and plunged nose first toward the ground. Impacting several seconds later, it threw up a mass of dirt and flame, leaving a 40-foot crater, even without an explosive warhead.

Since the beginning of 1944, General Metz had been familiarizing himself with the intricacies of the V-2 in his new position as LXV Corps tactical commander for the field employment of the weapon. Metz had visited the V-2 training school at Köslin and was satisfied with the curriculum and the enthusiasm shown by the director, Colonel Stegmaier. He had become convinced that ordinary soldiers could be trained to sufficiently to prepare and fire the missiles. However, he thought there was a general lack of coordination between the duties of the men of the firing batteries, which had been intentionally imposed months earlier by Dornberger for security reasons. He visited the troops at Grossborn and Heidelager and found Battalion 836 to be the only unit close to operational readiness. Battalion 485 was not up to strength in either men or equipment, and their field training had just begun. Their field trials exhibited an elementary state of training in many areas and required greater centralized control. Specialized vehicles were being manufactured at a rate that allowed for only one battalion to be equipped every six to seven weeks. Production at the Mittelwerk was moving at a slower-than-expected pace. Even though the number of rockets coming from the assembly lines had almost doubled over the past few months, with about 170 rockets invoiced for March, this was behind the anticipated schedule of 900 per month. Moreover, mechanical problems with the missiles continued at Blizna. By the middle of March, the success rate at Heidelager was dismal: out of 57 rockets launched, only 26 got into the air. Of those 26, only seven impacted, with only four falling in the designated target area. The cause of the airbursts was still unknown.

Following his tour of the installations, General Metz's conclusions were not favorable. In a report to LXV Corps headquarters, he stated that it was doubtful whether more than two under-strength battalions could be ready for operations by the end of July. He stated it was unacceptable for a divisional commander such as himself to command such as small, ill-trained body of troops; the V-2 battalions should be under the command of a colonel, not a general. He ended his report with a request for transfer to another command, but this was denied.[24]

On May 18–20, 1944, there was a V-2 demonstration at Heidelager. General Metz was present, and at the conclusion, it was apparent that another few months would be necessary before V-2 operations could begin. LXV Corps drew up plans for the V-2 opening attack, which would be code-named "Operation Penguin," tentatively scheduled for September of 1944, while the V-1 opening offensive was planned for mid-June. The mobile strategy was agreed upon, involving only one battalion, as it was doubtful that all units would be ready by September. This gave General Metz another excuse to request his relief from such a small command. He believed it was ludicrous that a general officer should be required to coordinate between corps headquarters and a single battalion. However, he was persuaded to withdraw his request by General Buhle, who was Army Chief of Staff, after Buhle promised that in time there would be a total of four complete V-2 battalions under two regimental staffs.[25]

Metz was not the only disgruntled commander. On May 31, 1944, Dornberger sent a letter to General Fromm requesting once more that he be given full authority over the V-2 project, including research and field operation. He went so far in his memorandum as to threaten General Fromm that he would apply to Hitler if necessary if his request was not met.[26]

While at Grossborn, Battalion 836 had been split between the northern camp, Grossborn-Linde, and the southern camp, Grossborn-Westfalenhof, with the practice ground near the railway line. Practice maneuvers that involved erecting the rocket into the firing position were carried out at night under the cover of darkness. It was rumored that the unit's activities might be attracting the attention of the Polish underground. In May of 1944, following the arrest of several suspected spies, Battalion 836 loaded all of their equipment, including specialized vehicles, onto a train bound for western Germany. A new practice ground was established at Baumholder in Rhineland-Palatinate, away from the prying eyes of enemy agents. By June, the first battery of Battalion 836 was quartered in the village of Wieselbach, while the second and third batteries were billeted in the villages of Mambächel and Erzweiler, respectively.[27]

Just before D-Day, London began receiving intelligence reports of the German rocket activity at Blizna. Recent Allied advances in Italy made it possible for the RAF to launch reconnaissance flights over occupied territory that had previously been out of range. On April 15, an RAF Mosquito snapped the first photographs of the SS compound at Blizna, but the pictures brought no conclusive evidence of rocket activity. On May 5, cameras caught a rocket in the open. The photograph revealed the rocket on a large transport trailer. Comprehensive reconnaissance flights confirmed the rockets could be transported not only by railway but also by road, which would mean they could be launched from just about anywhere.

While authorities in Britain were hunting for clues, Polish citizens near the V-2 test range at Heidelager had a front-row seat to the trials at Blizna. Members of the Polish underground, alerted by the thunderous noises in the distance, had witnessed strange objects flying into the heavens ever since December 5, 1943. On the evening of January 29, 1944, they heard an explosion inside the Heidelager camp, an apparent malfunction. The Polish resistance set up a chain of small units responsible for racing the Germans to the impact sites of the test rockets. On May 20, 1944, a V-2 launched from Heidelager at Blizna landed in a marsh close to the bank of the River Bug near village of Sarnaki. Members of the resistance found it before Germans arrived and pushed it into deep water, hiding it from view. A local farmer led a herd of cattle into the river, which stirred up the muddy water, further shielding the rocket. That night they brought six horses, pulled the rocket remains out of the mud, and hid it in an old barn nearby. A few days later, a team of Polish engineers dismantled most of the vital parts and packed them into barrels. It was then that Operation Trzeci Most (Third Bridge) was put into action. British Special Operations Executive, SOE, undertook this operation in cooperation with the Polish resistance to smuggle pieces of the V-2 out of occupied Poland for further study.

On July 25, 1944, an RAF C-47 Dakota took off from Brindisi, Italy, at 7:30 PM escorted by a Polish B-24 Liberator. On board were two pilots and four passengers. After nightfall the Dakota crossed the Yugoslavian coastline. It was at this point that the escort aircraft turned back for Italy, leaving the Dakota on its own. The twin-engine aircraft flew alone through the night, crossing Yugoslavia, Hungary, Czechoslovakia, and Poland. The pilots located their orientation point for landing by following the egress of the River Dunajec to the River Vistula. Right on time, the Dakota exchanged signals with the Polish party waiting on the ground, and soon lights appeared on the four corners of the remote landing strip. The pilot made two passes over the area and landed on the third attempt. The receiving party greeted them. The four passengers transferred equipment and moved out of the aircraft while the pilots remained. Five new passengers boarded carrying the smuggled V-2 parts and written reports. In a matter of minutes, the aircraft was ready for takeoff, but they quickly discovered the Dakota had become stuck in the soggy field. After an hour of frantically trying to free the aircraft, they finally succeeded, and the wheels of the Dakota rolled down the runway in the dark. The aircraft returned to Allied territory with the V-2 parts onboard, arriving in the early morning of July 26.

The British not only returned with valuable V-2 components but also were given detailed papers transcribed by Polish underground scientists. These reports, along with many photographs, gave a general description of many internal parts, including parts that could not be carried back by the aircraft. There were drawings of the launch sites and buildings at Blizna and papers listing number of launches and information about the impact areas.

By January of 1944, the secondary construction was completed on the liquid oxygen factory inside the Blockhaus at Eperlecques. From the beginning of February to the end of April, the U.S. Eighth Air Force mounted no fewer than 11 separate air raids on the facility. On February 2, 1944, 27 B-24 heavy bombers of the 392nd Bombardment Squadron rumbled down a remote runway in Britain around 11:00 AM. They formed up behind aircraft of the 578th Squadron and soon rendezvoused with other aircraft of the 14th Combat Wing. Their target that day was the Eperlecques bunker near Watten. In the middle of the afternoon, 49 bombers, bombing on flares, dropped a total of 187 junior blockbusters (2,000-pound bombs) on the bunker. The American aircrews reported a number of hits with the loss of one bomber to antiaircraft fire. In reality, very little damage was inflicted on the bunker. More than 2,000 bombs of varying size were dropped by B-17 and B-24 heavy bombers. The forests surrounding the bunker were laid waste; the terrain was transformed into a lunarlike landscape. However, the bombers had lost the technological battle with the Blockhaus and its five-meter-thick roof. The Germans continued construction from the inside, protected by the thick, reinforced concrete, and installed three compressors for producing liquid oxygen. The 1,000- and 2,000-pound bombs were having no effect on the building but were destroying the rail and road networks.

The V-1 (also known as a flying bomb, buzz bomb, or doodlebug) was a pilotless aircraft that was powered by a pulse-jet motor and carried a one-ton warhead. V-1s were launched by a steam catapult from a fixed ramp and traveled at about 350 miles per hour at a height of between 2,000 to 3,000 feet and had a range of about 150 miles. It was a nominal 27 feet long and had a wingspan of only about 18 feet. On June 12, 1944, the first V-1s were launched against London from ramps in the Pas de Calais on the northern coast of France. Over the next few months, thousands rained down on southeast Britain. The attacks created panic in London, and during the summer of 1944, 1.5 million people were evacuated from the city. The V-1's straight and level path made it a relatively easy target for the new automated antiaircraft gun systems, featuring the newly introduced radio proximity fuse, which allowed a shell to explode when it came to within a certain radius of a target. As Allied gun crews became more experienced, the number of V-1 kills rose dramatically. By August of 1944, the Allies were overrunning launch sites in the Pas de Calais, and the number of flying bomb attacks dropped dramatically. However, the German V-1 crews built new launching sites in Holland and directed their fire against the harbor at Antwerp. (*NARA*)

Near Saint Omer, construction on the massive concrete dome of the Wizernes bunker had begun in November of 1943. The Allies were aware of the German construction project but were slow in determining its purpose and even slower in targeting it for air raids. The building site first came under attack on March 11, 1944, but the raid proved to be very ineffective. The bombing attacks wreaked havoc in the nearby villages, upsetting the road and rail networks, but the concrete cupola remained strong and intact. Work on the project continued under the dome at a fast pace, even through some 299 air-raid warnings. On April 27, 1944, 16 American aircraft dropped 128 bombs on Wizernes, each weighing 1,000 pounds. On June 22, 1944, the U.S. 303 Bombardment Squadron flew a morning mission to Wizernes with a small force of fourteen B-17s. The mission was foiled due to a heavy cloud cover. The bombs fell short of the bunker in a wooded area east of the target, with one B-17 being shot down by antiaircraft guns. In total, 16 raids were carried out against Wizernes by Allied air forces. Allied pilots reported heavy, accurate flak batteries in the bunker area.

The Allied invasion of the European continent through Normandy began during

the early morning hours of June 6, 1944. The D-Day plan, known as Operation
Overlord, had been in preparation since 1943. Supreme command was entrusted to
American General Dwight Eisenhower. Just after midnight on June 6, British and
American airborne forces landed behind the German coastal fortifications. They were
followed after daybreak by an armada of seaborne troops consisting of American,
British, and Canadian armies. Some 4,000 transports, 800 warships, and innumerable
small craft participated in the invasion, with more than 11,000 aircraft supporting
them from above. While naval guns and Allied bombers assaulted the beach fortifications, the men swarmed ashore. At the base of the Cotentin peninsula, the U.S. forces
established two beachheads: one at Utah Beach and the other at Omaha Beach. British
and Canadian troops landed near Bayeux and established three beachheads: Gold, Juno,
and Sword. The Americans were slowed by fierce resistance, while the British and
Canadians advanced quickly but were stopped before reaching Caen. In less than a
week, the beachheads at Normandy were joined, and the Allied foothold was complete.

Allied invasion planners were relieved that D-Day had passed without any flying
bomb strikes, but their sanguinity was short lived. Late in the evening on June 6, LXV
Corps headquarters ordered the V-1 troops in the field to begin the assembly of their
prefabricated catapults. Originally slated for June 20, the V-1 offensive, because of the
Normandy invasion, would now begin on the night of June 12–13, 1944. On June 10,
Belgian agents reported spotting a train with 33 wagons full of V-1s roaring through
Ghent heading for France. On the evening of June 12, as the appointed hour
approached, the crews were desperately trying to complete the assembly of the modular ramps and associated equipment. Finally, in the early morning hours of June 13, the
first flying bombs were launched. Britain received its introduction to this cruel little
Nazi surprise. Cruising across the Channel in the darkness, the first bomb came down
at 4:13 AM in a field at Swanscombe near Gravesend, many miles from its intended
target. In all, the Germans managed to launch only ten V-1s from seven completed
ramps that night, and only one bomb fell in London. It was not the spectacular debut
everyone had expected. LXV Corps soon halted all operations until the other 57 ramps
were completed.[28]

On the night of June 15, 1944, the V-1 "blitz" of London and Southampton inten-
sified. Hundreds of flying bombs were catapulted into the darkness. On the evening of
the following day, the V-1 attacks were announced to the British public. It was plainly obvious to the citizens of London that they were under attack. More than 200 flying bombs had rained down on southern England, with 73 hitting London. The
onslaught was not without problems for the Germans. Four dozen V-1s had crashed
during or shortly after launch, and nine launching ramps were mangled in the process.
A few days later, a procession of flying bombs came over day and night; for two weeks
the attack continued at the rate of about 100 flying bombs per day. In one of the worst
explosions, a railway bridge was hit in east London causing extensive damage. Six died
in the explosion and 200 were left homeless. It was clear that the German V-1 crews
had quickly recovered from the blows dealt to them by Allied air power in the months
prior to the Normandy landings.

The V-1s carried a 1,900-pound warhead to a maximum range of about 250 miles. The V-1's ramjet engine used ordinary gasoline and made a loud buzzing sound—some said, like a fractured car muffler—so they were soon commonly called "buzz bombs" or "doodlebugs" by the British public. As the V-1 approached its target, the buzzing would suddenly stop, and the bomb would then fall silently to the ground and explode. Civilians began to realize they were safe until they heard the engine cut off, and then it was time to dive for cover. The V-1s were particularly terrifying because they would arrive at all times of the day and in all types of weather. The V-1 was not controlled from the ground after launch, but instead was directed to its target by a

On June 30, 1944, as seen from Fleet Street, London, a V-1 crashed in Aldwych nearby, killing 25 persons. Smoke of the explosion rises above St. Dunstan's. (*Daily Mirror*)

simple guidance system: a gyroscope system driven by compressed air to keep the missile stable, a magnetic compass to control bearing, and barometric altimeter to control altitude. When the guidance system determined that the rocket was at the target, the control surfaces shifted to put the bomb into a steep dive to its target, stalling out the engine in the process.

Hitler's long awaited V-bomb offensive had started. No one was more elated than the Führer himself. The combined American and British bombing offensive against German targets had taken its toll, and now he could strike back. On June 17, 1944, he flew to France to congratulate his generals for opening the offensive. General Heinemann suggested to Hitler that if the V-1 was to alter the course of the war in any fashion, production of the flying bomb would have to be stepped up. Many more V-1s would need to be fired. Adolf Hitler clearly expressed his intent to bombard London from launch sites in northern France with possibly as many as 3,000 missiles a day until the British capital was reduced to rubble.

On Sunday, June 18, 1944, during the middle of a Sunday morning service, a V-1 struck the Guards Chapel at Wellington Barracks, very close to Parliament Square and Whitehall. The church was totally demolished, and 121 worshipers lost their lives. By the end of June, the attacks had killed more than 1,700 people, leaving many more injured. The effect of a bomb falling so near to Whitehall prompted Crossbow officials to step up their attacks on the launching sites.[29]

While the V-1 was raining terror on London, the promoters of the V-2 were still struggling with the breakup dilemma. Hitler was losing patience with the rocket and wondered if it would ever be ready. On July 6, 1944, he ordered that another underground factory that was being excavated by slave laborers at Ebensee in Austria, code-named Zement (Cement), should be given to tank production. Following the bombing of Peenemünde in 1943, the new site at Ebensee had been planned for the relocation of the rocket center's development personnel.

Following the invasion of Western Europe, opposition to Hitler increased as Germany's military situation deteriorated. A plot was hatched by some in the German Army to assassinate Hitler and move for peace with the Allies. The leaders of the plot included retired Colonel General Ludwig Beck, Major General Henning von Tresckow, Colonel General Friedrich Olbricht, and several other top officers. The most stalwart conspirator was one Colonel Claus von Stauffenberg, who personally carried out the assassination attempt. On July 20, 1944, von Stauffenberg left a bomb in a briefcase in a conference room at the Wolf's Lair headquarters at Rastenburg, where Hitler was meeting with his top military aides. After arming the bomb, von Stauffenberg slipped out of the meeting. He witnessed the large explosion from outside, and, convinced that Hitler must have been killed, flew to Berlin to join the other plotters. The scheme called for the seizure of the Supreme Command Headquarters, but indecisiveness on the part of his coconspirators thwarted the plans. They failed to act until von Stauffenberg landed near Berlin more than three hours later. What von Stauffenberg did not know upon his arrival was that Hitler escaped the explosion with only minor injury. By then it was too late. Rumors of Hitler's survival melted the resolve of many of the key officers. At Berlin headquarters, General Fromm, who had known all about the plot, sought to prove his allegiance to Hitler by arresting a few of the chief conspirators. Von Stauffenberg, Olbricht, and two aides were promptly shot in the courtyard in the back of the War Ministry before they could implicate Fromm.

The next morning, General Dornberger traveled to the War Ministry to make a routine report to General Fromm. When he arrived, he had no idea what was going on. Dornberger was surprised to see the bodies of the conspirators, piled in a heap, near one of the doors. It was obvious the men had been shot. He recognized one of the bodies as that of General Olbricht, although his shoulder badge had been ripped from his uniform. Shortly afterward, the bodies were taken away. Entering the building, Dornberger was informed of General Fromm's absence. He decided to wait a few hours because of the circumstances. Reichsführer SS Himmler arrived later in the morning to address the nation.[30]

The attempt on the Führer's life stirred fear and uncertainty throughout the German population. In subsequent days, the Gestapo rounded up the remaining conspirators, many of whom were tortured to reveal their accomplices and hauled before the Volksgericht (People's Court). About 180 to 200 plotters were shot, hanged, or brutally strangled with piano wire. Hitler, in a move meant to humiliate the Army officer corps, granted the SS leaders military commands. Even Fromm was eventually arrested, tried, and executed. Himmler took over Fromm's position as Commander in Chief of the Reserve Army, a move that finally gave the SS total control over all aspects of the V-weapons. With Army leadership for all intents and purposes dissolved, Kammler was ready to pounce.[31]

By the end of July, the V-1 attacks were in full swing. Each day hundreds of flying bombs were being launched at London. On August 2, 1944, the Germans launched 360 V-1s, of which 107 were able to penetrate the defense zone around the city. Living up to Hitler's expectations, they were indeed causing terror for the population.

While Britain endured the flying bomb onslaught, the hunt for intelligence about the V-2 continued. On June 13, 1944, near the eastern coast of Sweden, residents in the small village of Bäckebo heard a thunderous boom and witnessed a shower of metal parts falling from the sky. Only minutes before, a V-2 equipped with an experimental guidance system, one that was destined for integration into the new Wasserfall antiaircraft missile, had been launched from Test Stand VII at Peenemünde. The operator of the remote guidance system had lost control of the rocket as it disappeared into the clouds. The German Embassy immediately contacted the Swedish government and demanded the return of the rocket debris. The Swedes logged a protest and refused the return of the fragments. Although Sweden was officially neutral during the war, a report of the incident was given to the British government, and later, British experts were allowed access to the wreckage and studied it in detail. Simultaneously, more rocket reports and sample parts began to be delivered from the resistance in Poland.

While struggling with the last-minute problems of the V-2, Peenemünde was also in the middle of another missile project. By August of 1944, over 1,000 employees were working on the Wasserfall antiaircraft missile program. The Wasserfall was essentially a small-scale development of the V-2, sharing the same general layout and aerodynamic shape. Since the missile was required to fly only as high as the attacking bombers, it could be much smaller than the V-2. The Wasserfall also included an additional set of fins located at the middle of the fuselage to provide extra control while maneuvering. Unlike the V-2, Wasserfall was designed to stand ready for periods of up to a month and fire on command, therefore, liquid oxygen could not be used for fuel. A new engine design, developed earlier by Dr. Thiel, was based on Visol (vinyl isobutyl ether) and SV-Stoff, or Salbei (nitric acid and sulfuric acid), a hypergolic mixture that was forced into the combustion chamber by pressurizing the fuel tanks with nitrogen gas. Guidance for use against daytime targets would be directed by radio control, while at night, a more complex system would be required because of low visibility. A new system known as Rheinland was under development for this. Rheinland used a transponder in the missile for locating it in flight by means of a radio direction finder and radar on the ground. A simple mechanical computer guided the missile into the tracking radar beam as soon as possible after launch, at which point the operator could see both the missile and the target on a single display, and then guide the missile onto the target. The Wasserfall was the second-most important project at Peenemünde during 1944–1945.[32]

Following the Normandy landings, there was an emergency meeting on June 8 in Berlin to discuss the future of unfinished V-weapon construction projects in France.

The V-2 bunker at Sottevast, which was situated south of Cherbourg on the Cotentin peninsula, was essentially already lost and would be overrun by advancing Allied armies in a matter of weeks. Priority was given for the completion the dome-covered bunker at Wizernes, which was now the only remaining V-2 launching bunker, and, incidentally, one of Hitler's favorite projects. Manpower was increased, but because of the nonstop bombing, a completion date could not be determined. However, Dornberger felt further construction at Wizernes was pointless. He was now more convinced than ever that the V-2 should be deployed as a mobile weapon and that the static bunker concept was pure folly. His representative at the meeting in Berlin had asked General von Rundstedt to consider abandoning the construction site at Wizernes altogether.[33]

Meanwhile, the British had been designing much heavier bombs such as the "Tallboy," or earthquake, bombs, each weighing 12,000 pounds. The RAF launched two attacks on the bunker at Eperlecques using the monstrous bombs. On June 19, 1944, 17 RAF Lancasters each dropped a single Tallboy bomb on the site, and during the second attack on July 25, 1944, 15 more Tallboys were dropped. Of the 32 bombs, only one actually hit the bunker. Falling on the northern lip of the roof, the gigantic bomb was only able to slightly pierce the shell of the roof. The impact caused hardly any damage to the structure but shook the building violently. Another Tallboy bomb hit the ground 27 meters from the south side, churning up the earth and creating a voluminous crater. The huge bombs, though never piercing the bunker, would cause violent mini-earthquakes each time they exploded. This prompted the German engineers to remove the liquid oxygen compressors for fear that they might explode under such conditions. On July 18, 1944, Hitler ordered the abandonment of the bunker. In the end, the Crossbow bombing campaign, along with the swift advance of Allied armies across northern France finally won the struggle against the Blockhaus at Eperlecques. The Eperlecques site was captured by Canadian troops on September 6, 1944, just a few days before the beginning of the mobile V-2 operations.

Previous Allied air raids on the dome at Wizernes were seemingly ineffective. The massive dome covering the underground construction appeared to be impregnable. It was not until July 17, 1944, that an attack by the RAF was somewhat successful. Sixteen Lancaster bombers each dropped one of the new Tallboy bombs on the bunker site. Although the bombs could not penetrate the dome, three of the Tallboys exploded next to the vertical exit tunnels. One burst just under the dome, and another burst in the mouth of one tunnel. The whole hillside collapsed, undermining the dome support, and covering up the two rocket vertical rocket exit ways.

Upon examining the damage following the raid, General Dornberger's staff reported that although the construction itself remained largely intact, the earth surrounding the Wizernes bunker was so churned up, there was a possibility that the dome might collapse. Dornberger recommended to LXV Corps on July 28, 1944, that the site should be abandoned. The Todt engineers disputed this finding, but the Allied invasion of France, along with dwindling supplies, prevented any further construction. American forces under General Omar Bradley had cut off the Cotentin peninsula on

June 18, and Cherbourg surrendered on June 27. Now, virtually all V-2 positions west of the Somme were no longer valid. Shortly thereafter, General Eisenhower and General Bradley toured the massive excavation site for the V-2 bunker at Sottevast. At the end of July 1944, Hitler ordered that Wizernes should also be abandoned, along with its guidance bunker at Roquetoire. The 500 Russian prisoners working on the construction site were dispatched by train to Germany and were never seen again. Thus, the Allies had won the battle with the bunkers. Not a single V-2 would ever be launched from these concrete monsters on the coast of France. The missile would now be deployed exclusively as a mobile weapon.

At age 16, Heinz Wellmann received his conscription order by mail and was immediately required to report to the Reichsarbeitsdienst (state labor service). Just after he turned 17 years old, he left his small village near Minkowsky and traveled to Namslau. Soon after, he, along with other recruits, was moved to Brieg. While there, some SS officials came in and asked his group, "Which of you will volunteer for the SS?" When no one came forward, they said, "OK, then we will make volunteers." Of the several hundred young men present, the SS took all but a few. Young Heinz was transported to Nuremburg where he received basic combat training and instruction in radio communication and telecommunications. When the training program was completed, Heinz received word he was to be assigned to the eastern front. Around 180 of the trainees were already geared up and ready to go, when suddenly some SS men came in and selected a dozen of the soldiers, including Heinz, for some type of special assignment. They were pulled aside and told nothing. From Nuremburg they were transported to SS headquarters in Berlin. After spending the night in a soldier's hostel, they received orders to report to a place called Nordhausen the next morning.

In the spring of 1944, Heinz and his fellow soldiers arrived at Nordhausen. They were not given any clue as to why they were there, and no one knew anything about the underground plant nearby. They were housed in newly constructed barracks, just beyond the perimeter of the plant. The barracks were comfortable, though the ground was beginning to thaw from winter and mud was everywhere. During the next few days, they were joined by more and more soldiers, and soon the barracks were full. The men came from everywhere, each one of them being transferred from some other duty post; all were unaware of the reason for their transfer. The answer came soon enough when it was announced they had been selected for an exclusive SS long-range rocket battery. They were in Nordhausen to receive specialized training at the rocket factory. Heinz and his fellow soldiers would be trained to launch the new Wunderwaffe, the V-2.

Prior to entering the factory, they were given lessons dealing with the basics of rocket theory. They were shown photographs and silent movies of the rockets, which gave them a general idea of their future mission. After this initial training, they were divided into several groups of 20 and told the remainder of their training would be conducted inside the factory. They were issued special overalls, to appear as civilians, and were told not to identify themselves as SS soldiers while their training was being conducted inside the tunnels. No military protocol was to be observed, not even a salute to a superior officer. As each group entered the tunnels in rotating shifts, they received a detailed technical overview of the rocket. The construction techniques of the apparatus were explained. Each component was shown and commentary was given to explain its function. However, nothing was hands on, they only watched as the forced laborers assembled the rockets. It was forbidden to talk to the concentration camp prisoners, but Heinz and the others did anyway. The conversations were short because the prisoners were wary of being caught by the German guards.

As the lessons moved to various locations throughout the factory, the trainees gained a real sense of the immense size of the underground factory. They watched as whole freight trains entered the mountain through the main gallery. The rockets were loaded for transport on the railway cars and covered with canvas camouflage. Leading from the main tunnel were many tributary galleries, each pertaining to various components. A second main tunnel was still under construction, and everywhere they saw a flurry of activity. After each training session, the soldiers were required to sign a document stating they understood that everything was "top secret" and it was not to be discussed outside of the factory. They could not leave the tunnels with any notes or papers and were required to commit each lesson to memory. It was all very bewildering for young Heinz.[34]

By June of 1944, over 1,000 rockets had been produced and invoiced at the Mittelwerk. Of these, over 300 were defective and had to be returned for repairs. With over 17,000 individual components, there were constant modifications incorporated into the rockets coming off the line. Rockets with differing warhead shapes were fired by the Training and Experimental Battery 444 at Heidelager. The development of a sensitive fuse, one that could trigger quickly enough before the rocket could bury itself in the ground, was now the focus of the engineers. However, it was a problem that would never be solved. The warhead had to be converted to a six-millimeter-thick steel casing because of an extreme shortage in Germany of the aluminum-magnesium alloy that was used previously. The amount of explosives in the warhead had to be reduced moderately to maintain the overall range of the missile.[35]

On July 24, 1944, the final V-2 launch from Blizna soared in the heavens. A few days earlier, the camp had been attacked by Polish partisans, who were quickly repulsed by German troops. However, this gave the rocket crews only enough time to conduct last-minute tests and then remove essential vehicles and equipment. Due to the advance of the Red Army, which was closing in from the east, Heidelager had to be evacuated.

Since the beginning of the tests at Blizna, General Kammler made himself a frequent visitor to the test range. He wasn't expected sometimes or even welcomed, but he imposed himself as Himmler's eyes and ears. He mingled among the men, asking questions as if genuinely interested. Later, he would use the information to his advantage by highlighting differing opinions and causing instability in the ranks. Some saw it for what it was. Kammler's lack of character was evident to many when he began referring to von Brown as an arrogant youngster, too cavalier and full of himself. He thought of Degenkolb as only a drunkard. He went so far as to call for Dornberger's court-martial on the grounds that Germany's military capacity had been weakened by the rocket team's "frivolous" endeavors. Kammler no doubt saw Dornberger as a rival. The two men argued over the airbursts and other technical problems delaying the deployment.[36]

After difficult fighting, the American forces in Normandy had finally broken out of the hedgerow country and on July 18, 1944, captured the vital communications center of Saint-Lô, cutting off the German forces under Field Marshal Erwin Rommel. That same day, as the summer sun beat down, technicians scurried about Test Stand VII in preparation of another test firing. The activities came to a halt when a telephone call reported the approach of enemy aircraft. People headed for shelters without hesitation, and soon the air raid sirens wailed. While the battle raged in Normandy, Allied air commanders had not forgotten about the research establishment at Peenemünde. The American intelligence community had developed serious suspicions that the top-secret German complex on the Baltic may well have been conducting other types of experimentation, possibly in the area of nuclear weaponry. Under the arena walls at Test Stand VII, the technicians closed the steel doors to the control room and waited. The sound in the distance became louder and louder as 379 heavy bombers of the U.S. Eighth Air Force approached. The earth began shaking as the bombers unloaded 920 tons of bombs on eight different targets around the peninsula. When it was over, the workers emerged from their hiding places and surveyed the damage. Again, as in August of the previous year, many of the bombs had fallen in the woods. But there was a considerable amount of damage at P-7—the control room had been hit, almost crushing the occupants; the water tanks and pump station were demolished; and one the mobile test stands was completely ruined. Mangled rockets stood in the battered assembly building, and just beyond, craters walked right across the area of Test Stand X.[37]

At the end of July 1944, a hasty evacuation from Blizna was underway, and a new rocket test location was established in the Tuchola woods near Wierzchucin and Cekcyn, about 160 miles northwest of Warsaw. The new firing range, code-named Heidekraut (Heather), was approximately ten miles east of the town of Tuchel, in the area east of Wierzchucin at Suchom Lake. On August 2, 1944, a rocket was launched from a platform constructed of cut logs placed in the ground covered by a steel plate near the banks of the lake. Soon, the Polish village of Lisiny awoke to the thunderous noise of a V-2 rocket engine. A few seconds later, the awful noise was transposed into a tremendous earthquake. In the village, walls burst and ceilings collapsed. The explosion uprooted trees and created a 20-foot-deep crater in the middle of an orchard with rocket pieces burning at the bottom and the whole area smelling of alcohol. Minutes after the explosion, German troops, along with a doctor, came to the village to help rescue and treat the injured. Later, a Feuerleitpanzer halftrack appeared and carried away various pieces of the shattered missile. The next day the Germans evacuated the whole village.

Rockets were brought by train to the railway station in Wierzchucin, and then on a side track, the rockets were loaded on long Vidalwagen trailers and dispatched into the forest in the direction of Suchom Lake. Shipment from Wierzchucin eventually proved to be troublesome; later the rockets were transported to station Lisiny, where a convenient loading ramp already existed, constructed by the Polish Army prior to the war. Completing their training at Heidekraut were crews of the SS Werfer Battery 500 and the first battery of Battalion 836. Later, they would be joined by the newly formed second battery of the Battalion 485.

It was forbidden for civilians to walk near the V-2 operational areas, and roads were blocked with barriers guarded by German patrols. Eyewitnesses reported that some of the rockets malfunctioned and exploded in mid-air, falling into the nearby lake, while others flew in the direction of Toru?. German recovery teams were observed by civilians attempting to salvage malfunctioning rockets from the water. The target areas for rockets launched from Heidekraut were approximately 152 miles to the south, in the areas of Lutatow and Lodz. It is interesting to note that in the spring of 1944, SS authorities planned the deportation and extermination of the more than 160,000 Jews in the Lodz Ghetto, rendering any potential mishaps in the target area inconsequential in the minds of the SS.[38]

Throughout the summer, Allied warplanes continued their sorties to locate the extremely well hidden V-1 launching ramps in the forested areas of Pas de Calais. On June 20 and again on June 29, the Americans attacked the Volkswagenwerk at Fallersleben in an attempt to destroy the V-1s at the source, but the raids failed to reduce the number of flying bomb attacks. Nevertheless, during the first week of July 1944, Crossbow raids caused significant destruction to the fixed sites. On July 4, the mammoth concrete V-1 bunker at Siracourt, which had been bombed 30 times since April, collapsed after receiving a direct hit from a Tallboy bomb. The vast Allied effort reached its peak around the end of August, but in the meantime, the death toll mounted in Britain. Fortunately for Londoners, events in northern France would bring some relief. By mid-August, British forces from Falaise in the north and American forces moving from Argentan in the south closed the gap of the "Falaise pocket," wiping out most of the German Seventh Army. The Germans desperately retreated toward the Seine River while being chased by the U.S. Third Army under the command of General George S. Patton. Fearing a second encirclement west of the Seine River, the Germans fought to save their dwindling escape routes. By the end of August, the U.S. Third Army had raced all the way to the Saar River while the British and Canadians moved north. The Allied push in northern France ultimately brought a temporary halt to the V-1 attacks, as the launching ramps were overrun by advancing Allied troops.

The death of Test Stand VII occurred on August 4, 1944, when American bombers once again attacked Peenemünde. Improving their accuracy, the bombardiers from 221 aircraft walked a carpet of explosives right across P-7. However, it was not the last time Peenemünders would hear the air-raid sirens. On August 25, 1944, the same day Paris was liberated by Allied armies, 146 bombers of the U.S. Eighth Air Force inundated the rocket establishment with bombs. It was the final nail in the coffin. Even though the attacking force was smaller, the damage was far worse than previous raids. There would be no more test launches from P-7; the whole arena was left in shambles.[39]

By the beginning of September, the Crossbow attacks had dropped nearly 118,000 tons of bombs on the V-weapons targets since the initial attack on Peenemünde more than a year earlier, of which 20,000 tons had been directed at locations suspected as being directly associated with the V-2 rocket. During the nearly 37,000 sorties flown against Crossbow targets, the Allies lost about 450 heavy bombers, along with 2,900 brave airmen, most of whom were British.[40]

After the fall of Paris, the office of HARKO 191 under General Metz was ordered to withdraw from France. All V-2 related personnel and equipment was evacuated by way of Ypres to Turnhout, then to Dongen near Tilburg en route to Germany. On August 29, 1944, there was a conference between General Jodl, General Buhle, and SS General Kammler at which tactical command of all troops was given to HARKO 191. Hitler had ordered the start of Operation Penguin to commence as soon as possible. When detailing the plans for the operation, Hitler decreed that half of the rockets should be aimed at Paris and half at London. On August 30, the General Staff issued orders to LXV Corps to begin the preparation of launch sites near Ghent for an impending rocket attack against Paris. The following day, LXV Corps requested proof of the Führer's order and asked General Keitel, because of Kammler's involvement, for confirmation of the current chain of command. Kammler assumed the confusion had been laid to rest when he informed commanders in the west that he had been given provisional supervision of the rocket project. Per Himmler's orders, the opening V-2 attacks would commence on September 5. He ended by stating that if anyone needed to get in touch with him, he was traveling to Brussels.[41]

A few days later, on August 31, 1944, the important players gathered in Brussels for what was believed to be the final planning meeting for the rocket offensive. SS General Kammler had called the meeting, but General Heinemann presided. Present were various Chiefs of Staff, including Colonel Walter, General Metz, and General Dornberger. In an attempt to subdue those present into concession, Kammler boldly announced that he was taking control of not only rocket production and administration but also tactical deployment, a move that did not sit well with LXV Corps commanders. Kammler began issuing commands for units in the field and detailing the operational strategy. A heated argument ensued, after which General Heinemann demanded to see his papers from the Armed Forces High Command. The SS general could produce no such papers. Then, Colonel Walter, LXV Corps Chief of Staff, proceeded to read a communiqué from the High Command that confirmed the authority of LXV Corps over tactical deployment of the rocket. The message went on to stress Kammler's complete irrelevance. The legitimacy of Kammler's claim collapsed and his power grab was averted, but not for long.

Two days later, Himmler confirmed General Kammler's complete authority. The Armed Forces High Command instructed LXV Corps to brief Kammler on the preparations for the offensive. At this point, LXV Corps would still conduct the operation, although Kammler would be in command. Himmler's appointment gave Kammler total control of the program, including operational deployment. He would be subservient to only Himmler and to SS General Jüttner, the head of the SS leadership office, which acted as general staff for the Waffen SS. To solidify his control, Kammler immediately launched a campaign to isolate Dornberger. He asserted the right to issue direct orders to Dornberger's Chief of Staff, Colonel Thom. No doubt Kammler's intentions were meant to cut the legs out from under his rival and force Dornberger out of the program completely.[42]

Even before Kammler's ascendancy, Dornberger was struggling to retain what little influence he still processed. He was no longer commander of Peenemünde, the Army Ordnance Department saw him now as an outsider, and LXV Corps had replaced him as commander of the rocket batteries. Kammler's ruthless acquisition of power over the rocket project left Dornberger in complete despair. It seemed he had lost everything. On August 29, 1944, he drafted his resignation only to be convinced by his Peenemünde colleagues to change his mind. Ironically, it was SS General Jüttner who ultimately rescued Dornberger from Kammler's malevolence. When Colonel Thom, on Kammler's orders, reported to Jüttner in Berlin to complain about Dornberger and request his removal and transfer of his staff, the reaction from Jüttner was probably not what Kammler expected. Jüttner was outraged at Kammler's treatment of Dornberger. He felt it was insufferable that Kammler had threatened to have Dornberger shot if he dared approach Jüttner himself. Considering Dornberger's background and knowledge of the rocket, Jüttner refused to expel him. He ordered the two men to work together for the good of the program. Dornberger would soon become General Kammler's representative on the home front, responsible for training the rocket troops, weapon improvements, and coordinating supply of materials to the firing areas.[43]

The opening of the rocket offensive was set for September 5, 1944, and per Kammler's request, the Armed Forces High Command issued the necessary orders to start the movement of the rocket battalions to the firing areas on September 2. Kammler put Dornberger in charge for getting the movement organized and underway. The next day, more than 5,000 personnel and approximately 1,500 vehicles were moving to the firing areas. General Dornberger had promised to have the V-2 batteries ready for action by September 15, 1944. These included the first and second batteries of Battalion 485, the second and third batteries of Battalion 836, and Training and Experimental Battery 444. The third battery of Battalion 485, along with the first battery of Battalion 836 and the SS Werfer Battery 500, were still in training and would not be ready for the start of the campaign.

On September 6, 1944, General Kammler issued his first command order. After reaffirming his authorization from Reichsführer SS Himmler, he ordered the troops divided into two groups: Gruppe Nord (Group North) and Gruppe Süd (Group South). Kammler dictated that the operational commanders for each group would answer to him personally and that no important decisions would be made without first gaining his approval. Group South, consisting of about 2,000 troops, was under the command of Major Weber, who had been in charge of Battery 444 since the beginning. His units consisted of Battalion 836 and Battery 444. Group North—about 3,000 troops—consisted of Battalion 485 (which was later joined by SS Werfer Battery 500), and was initially under the command of Major von Ploetz, although he would be replaced several weeks later by Major Schulz. Each of the firing batteries was appointed a designated technical battery and support groups. All units associated with the deployment of the rocket, over 8,000 men, were under direct command of SS Gruppenführer Hans Kammler, who had come up with his own name for his command: Division zur Vergeltung (the vengeance division) or Division z.V., which stood for Special Employment for Retaliation.[44]

By the last week in August, 1944, the Allies had crossed the Seine River, and the LXV Corps was withdrawing toward the Low Countries, having evacuated as much equipment as possible—most importantly, the machinery required to make liquid oxygen. The prepared launching sites selected by General Heinemann for the beginning of the operation, north of the Somme River, had to be abandoned. Instead, Kammler ordered Group North to the city of The Hague on the Dutch coast to begin the bombardment of London. Group South would be partially split up with Battalion 836 ordered to the area of Euskirchen near the German border to launch against the French city of Lille, while Battery 444 was given the special mission of bombarding Paris. Division z.V. headquarters was first established at Kleve and then moved a few days later to the small village of Berg en Dal, just over the Dutch border near the city of Nijmegen on the Waal River.

The rocket units were already concentrated in the Baumholder training area, 28 miles southwest of Trier, and they were immediately organized for fast deployment. The two batteries of Battalion 485, under the command of Colonel Hohmann, assembled near Kleve on September 5. From there they traveled across Holland to The Hague and set up operations in the influential suburb of Wassenaar. Battalion 836 moved from Baumholder to a point west of the Dutch border near Venlo, where they stopped and then withdrew to the area of Euskirchen in Nordrhein-Westfalen and were joined there by Battery 444 on September 3. The next day Battery 444, along with the first and second batteries of Battalion 836, moved into Belgium. Battery 444 was operational, but the units of Battalion 836 had yet to receive their launching equipment.

The men of Battery 444 were considered to be the most highly trained and proficient rocket troops available at the time. As the first organized launching battery, many members gained valuable experience from their time at Peenemünde or had spent months at Heidelager either firing test rockets or training others to do so. It was probably for this reason that Battery 444 was chosen for Kammler's special assignment of attacking Paris. He had ordered the rocket troops to fire on September 6. On September 4, 1944, the unit transported their rockets into the area of Plateau des Tailles, to a location south of Baraque de Fraiture on highway N15 at Petites-Tailles. The battery had not yet received their third firing table, so only two firing sites were prepared near the main road. In reality, Battery 444 had advanced too far and set up operations extremely close to the battlefront. The 2nd SS Panzer Division, Das Reich, was holding the line south of Battery 444's position, as elements of the U.S. 3rd Armored Division, just a short distance away, were pushing northeast. Belgian resistance fighters from the town of La Roche were on the roads ahead of the American advance, and on September 6, a group of about 30 partisans attacked Battery 444's parking area near Chabrehez, about 1.5 miles west of the launching positions. One officer of the battery was killed and another two were wounded.[45]

Earlier, Battery 444 firing crews had erected two rockets on their launching tables, and after fueling, the first round was set to go off at 9:00 in the morning. When the firing commander gave the signal for ignition, the engine came to life, and the flame below the rocket burned evenly. Then the turbine was engaged and the rocket lifted slowly off the firing table; immediately, the engine shut down and the rocket dropped back onto the table, each fin settling again on its perch. Having experienced many similar failures at Heidelager, the crew waited for the swaying rocket to topple over and explode. To their surprise it remained upright on the table. The technical officer ordered the missile defueled. About 40 minutes later, the second rocket was fueled and ready for launch. Amazingly, it did exactly the same thing. Both examples were taken down, and it was later discovered that faulty accelerometers were the cause of the premature engine stops. However, the loss of battery personnel, even before the first rocket had gotten off the ground, was disheartening for the men of Battery 444.[46]

For the Allies in Britain and on the continent, it seemed that the worst of Hitler's secret weapons campaign might soon be over. The threat of the flying bomb attacks had ended on September 1 after the last of the remaining catapults in Pas de Calais were captured. The V-1s had been beat, the V-2 bunkers were in ruin, and the prepared sites for the missiles were no longer available to the Germans. Newspapers in England were carrying headlines theorizing that V-2s might never fall on British soil. On September 4, 1944, British forces had rolled into Antwerp with no resistance, and the U.S. First Army was approaching the very border of Germany. People in Britain were wondering if the war might not soon be over.

But German forces still monopolized many strategic locations, including the seaward approaches to the port at Antwerp. The Allies would have to deal with these first, and because of their overstretched supply lines, they were hard pressed to rush into Holland and Germany. All supplies, including equipment, ammunition, food, and troops had to be moved by trucks from the Normandy beachhead, more than 450 miles across France. The advance had outrun the supply lines. Any further movement would have to be postponed until the deep-water port of Antwerp could be cleared and forward supply depots could be established.[47]

On September 3, 1944, all RAF commanders were ordered to suspend further attacks on suspected V-2 targets; the Americans had already ceased armed sorties against suspected launching sites on August 30. On September 5, the British Chiefs of Staff boldly announced that all V-1 and V-2 launching sites would be captured in a matter of days. On September 7, Duncan Sandys spoke to the press at Air Ministry headquarters about the battle against the German secret weapons. Walking up to the podium, he audaciously announced, "With the exception of a few parting shots, the Battle of London is over!"

What Sandys did not know was that the real terror was about to begin.

5

Operation Penguin

The missile, dressed in its ragged camouflage scheme, stood motionless on its sturdy firing table, hidden completely by the dense trees of the Ardennes. About 40 feet away, the measurement platoon had situated two optical aiming collimators that established the vertical positioning of the device. After the power supply vehicle had run a series of preliminary tests, the firing crew electricians had climbed the rungs of the Meillerwagen arm to install the 50-volt flight control battery, along with the two 27-volt batteries for the gyros. The fins of the rocket were situated in four pedestal slots attached to the rotating ring of the firing table. The crew used a hand crank to rotate the rocket on the table so that fin number one was aligned with the pitch axis toward Paris. The protective jet covers had been removed from the venturi in the combustion chamber, and the fragile carbon-graphite exhaust rudders were carefully bolted in place. It was a few minutes past 8 o'clock in the morning; the convoy of fueling vehicles would soon approach.

As a result of the two launch failures and because of the partisan attacks of a few days earlier, the firing crews of Battery 444 had moved from the launching position at Petites Tailles to a new position six miles to the east, near the tiny village of Sterpigny. On Friday, September 8, 1944, the first V-2 of the German rocket campaign lifted off at 8:40 AM heading toward Paris. While the firing crews must have been elated to finally get operations underway, indications are that the rocket exploded at high altitude, breaking up before hitting the target. However, two hours later, a second rocket fired by Battery 444 came down southeast of Paris, at Charentonneau à Maisons-Alfort, killing six people and injuring 36 others. It was the first ballistic missile attack in history, but it caused little commotion.

Because of the sweeping Allied advances just weeks earlier, many Dutch people in the province of South Holland were betting the war would be over by the end of the year. On September 5, 1944, there were many rumors about the fast-approaching Allied armies. However, the citizens of Holland were unaware of the overextended sup-

115

ply lines slowing the advance, which would soon become a crawl. The suffering for the Dutch was far from over. After several years of occupation by German forces, residents of The Hague (Den Haag, or officially, 's-Gravenhage), the administrative capital of the Netherlands, had become accustomed to the various sounds of war: the drone of aircraft, the blast of falling bombs, and the thud of antiaircraft guns; but now they were about to be introduced to a new sound, one that would soon bring trepidation.

Following an early morning storm, it was a warm, sunny afternoon in the affluent suburb of Wassenaar, just northeast of The Hague. On that day, September 8, 1944, the local residents heard a strange sound. A tremendous roar filled the air, furniture and windowpanes shook, and the ground began vibrating. Dutch civilians exited their homes in time to witness two pointed projectiles, about 50 feet long, rising slowly above the treetops. A cloud of smoke rolled through the streets after the rockets had cleared the trees. They gained speed rapidly with flames emerging at the rear extending to more than half their lengths. As the rockets moved higher and higher into the heavens, a white trail appeared in the form of a spiral. The smoke disappeared at a height of about 12 miles, and the projectiles accelerated rapidly, their trajectory flattening out somewhat. They finally disappeared traveling at a terrific speed. Under the command of Colonel Hohmann, the second battery of Battalion 485 had launched two rockets simultaneously at 6:36 PM from the picturesque neighborhood streets: one from the intersection of Lijsterlaan, Konijnenlaan, and Koekoekslaan; the other, about 150 feet away, from the intersection of Lijsterlaan and Schouwweg.

The firing crews had hoped to get an earlier start in the day, however, the strong winds from the passing storm had knocked down a tree, which fell on one of the radio tents in a position several miles away where the signal to shut down the rocket's engine in flight would be broadcast. Finally, late in the afternoon, German trucks and odd-looking vehicles were seen driving up to the Wassenaar sites from different directions. One of the vehicles was described by eyewitnesses as a long trailer having 16 wheels with some sort of lifting apparatus. Others included tanker trucks and trailers, which were filled from a railway wagon at Wassenaar Station, and an armored halftrack, which had been left about 1,000 feet away from the launching positions. Occupants of the nearby houses had been evicted the day before by German military guards who ordered them to leave their doors and windows unlocked; no one was allowed within half a mile of the area, but the noises and flurry of activity prompted a few daring souls to venture closer for a peek. The launching crews consisted of about 20 soldiers who, when fueling the missiles, were completely clad from head to toe with asbestos-like protective overalls and helmets. It was later discovered that the projectiles were fired using a power cable brought out from a nearby electrical supply point. The Germans had installed a number of these electrical cables, connected to normal power mains, which were laid in the neighborhood via the roads Rijksstraatweg and Rust En Vreugdlaan.[1]

Less than 200 miles away, in the West London suburb of Chiswick, people had already arrived home from their jobs. London was on double summer time, while

Holland was on daylight savings, so the hour was the same in both cities. Those who were still on the streets were in a hurry because of the increasing drizzle that threatened to become a steady downpour. Sixty-four-year-old Robert Stubbs was one who was running late. Stubbs had taken care of some last-minute chores before leaving his job as caretaker of the Staveley Road School. Crossing the school playing field, he noticed a serviceman moving briskly down the street past several houses. At 6:41 PM, without any warning, Stubbs was flung 20 feet across the field by a terrific blast. One of the rockets launched from Wassenaar had ended its five-minute journey traveling three times the speed of sound; it slammed into English soil at Staveley Road in Chiswick.

Double shot of V-2s launched from Hook of Holland or The Hague heading for London. This photo was taken from the outskirts of Antwerp looking north, almost 50 miles away from the firing locations. (Author)

A description of a V-2 impact would be as follows: first, a "whip cracking" sound of the blast wave was heard, created by the rocket moving faster than the speed of sound, which bounces off of the point of impact split seconds before the flash of impact; this was followed by the chaos of the explosion with debris and earth churned skyward. Immediately afterward, as if in reverse order, the whine and rush of whistling air was heard as the sound of the rocket descending through the heavens caught up with the rocket, followed by the roar of the incoming rocket, which tapered off to silence. There could be no warning.

After coming to his senses, Stubbs was shocked to see the devastation all around him. The bomb, which landed at the Burlington Lane end of Staveley Road, had demolished houses on both sides of the cherry-tree-lined street and left a 30-foot crater in the road. A 65-year-old woman, Mrs. Harrison, came crawling out of the shattered home at No. 3, covered in dust. Stubbs tottered over to help her, but there was little he could do; by the time help arrived, Mrs. Harrison had died. The explosion had killed three people, including Private Frank Browning, who was in a hurry that payday to visit his girlfriend's house. Seventeen other people had been seriously injured in the blast, covered by the debris of their wrecked homes until rescue workers pulled them free.[2]

Newspapers in Britain for September 8, 1944, came out with headlines proclaiming the end of the (V-1) assault on London. The articles featured photographs of

Duncan Sandys from his news conference the day before. Over at the Air Ministry Scientific Intelligence Office, Dr. R. V. Jones and his colleague Charles Frank had been discussing the validity of these optimistic calculations. During their conversation, there was a double bang in the distance; the two men looked at each other knowingly and then exclaimed almost simultaneously, "That's the first one!" The V-2, as if to scoff at the day's headlines, had finally made its appearance. Moments later, another sonic boom was felt; it was the second rocket from Wassenaar, which fell harmlessly in a field at Parndon Wood, some 16 miles north near Epping. The world had entered into a new type of warfare: the age of the ballistic missile.[3]

The thunderous impact at Chiswick was heard all across London. Homes shook, followed by some rumbling vibrations. People were puzzled; there had been no sound of a V-1 sputtering across the sky or the hum of a German bomber overhead or any air-raid sirens. Those living near the strike reported hearing a double bang, which was later discovered to be the boom of the missile breaking the sound barrier. People near the area filtered into the streets, curious to know what had just happened. Eleven homes situated on Staveley Road were totally flattened, while two dozen more sustained substantial damage. Many people were surprised to find the blast radius was so broad; windows of houses almost two miles away had been blown out by the concussion. Rescue workers moved up the street giving aid to those who were bleeding from cuts caused by flying debris and glass. Military and Civil Defense officials soon arrived at the scene and began rummaging about, collecting scraps of metal found near the crater. A reporter who was fast on the scene asked one of the Civil Defense men if the explosion seemed to be a new type of German guided bomb and was told, "We can't say what it was. It might have been a gas main explosion." The deep crater fashioned by the missile actually did rupture water and gas mains; however, the military officials on hand knew exactly what had happened—Big Ben (the British code name for the German rocket) had arrived.[4]

Immediately after firing the missiles, the soldiers of Battalion 485 had packed their equipment and driven away from the launching sites. Later that evening, the locals returned to their neighborhood. They found that both launching sites were in the middle of roadways passing through the wooded upper-class neighborhood. At each launching place, there was a circular patch from which the road had been melted or burned. The burned patch had a diameter of 20 to 25 feet, and in the center of each scorched area was an unburned area in a box form, suggesting that some sort of stand had been used. The beautiful large, mature trees near the edge of the roadway were very badly burned up to a height of about four feet and less badly burned near the treetops. There was also evidence of a violent low blast, because the grass and nearby gardens were flattened; all the leaves had strangely vanished from the ground below the trees. The thatched roof of a nearby small house had been lifted and blown off by the rocket exhaust. The following day, a report was sent to Kammler's headquarters: "The rocket weapon was effective. Two rounds launched against London."[5]

On September 9, 1944, on a tip from the Dutch resistance, RAF fighter aircraft were able to locate the neighborhood from where the first two V-2s had been fired. The pilots knew they were close when a heavy barrage of antiaircraft fire met them over Wassenaar. The aircraft bombed and strafed the neighborhood streets, but the rocket troops were no longer there. The second battery of Battalion 485 had set up operations on the grounds of the Beukenhorst estate, a few miles to the southwest; firing only one missile that day, which fell into the sea. The air attacks prompted local government officials to order the evacuation of Wassenaar on September 10, with the exception of the area behind the Kerkdam. Within three days, the entire area between the viaduct on the Leidschenstraatweg and Kerkdam, including the Marlot Park, was evacuated. On September 10, the first battery of Battalion 485 was ready to join the second battery in firing operations from the area. The mobile batteries, each with three firing tables, were dispersed to new firing sites in the Wassenaar and The Hague. On September 12, the soldiers moved into the Marlot area adjacent to the Duindigt horse racing track. That day, between the hours of 6:00 AM and 9:40 PM, five more V-2s were launched. Londoners experienced more of the mysterious bangs, one of which landed on the Chrysler manufacturing plant at Kew Gardens, killing eight people and causing extensive damage.

The amount of damage inflicted by both the V-1 flying bomb and the V-2 rocket at the immediate impact area exhibited total destruction. However, the collateral damage was greater during a V-2 strike. The crater made by a V-2 was much deeper than that of the flying bomb. Instead of knocking down walls and ceilings, the V-2's impact speed helped disintegrate the material into tiny pieces, throwing off some as lethal shards.[6]

On September 14, two rockets launched from Wassenaar failed; each one crashed into the North Sea, just off the coastline north of Kijkduin. Another launched from the grounds of the Beukenhorst estate hit Walthamstow, England, digging a crater 25 feet deep and killing seven people.

Nevertheless, the start of the V-2 bombardment shook the confidence of the British citizens. They knew their government was covering up or least obscuring the facts about the loud explosions. There was no official news released at first, and officials in London refused to acknowledge the mysterious blasts. Government officials did not create the rumor of the "gasometers" or "exploding gas mains," but did not deny them either. Yet, 16 gas main explosions in ten days was quite a lot; most people felt it had to be some sort of new nasty German surprise. Some people remembered earlier news columns reporting a second type of German robot bomb and quickly associated the new explosions with the previously mentioned V-2; mockingly referring to the new attacks as the "flying gas mains." Even though the V-2 was not completely unknown to Winston Churchill's War Cabinet, the opening of the rocket offensive had come as a shock to its members. It had been hoped that the raids on Peenemünde and the intensive bombardment of the Crossbow sites, along with the capture of the prepared firing sites in the Pas de Calais, would prevent the German missile attacks. But now, much to the dismay of the War Cabinet, the topic of conversation all over the city was focused on this new horror, with wild stories and rumors running rampant.

Just a few days earlier, the British populace, like the Dutch across the channel, had been expecting the war to be over soon. After all, hadn't the Allied armies driven ten miles inside of Germany already? Over a million people had been evacuated from London during the flying bomb offensive, and many were already making their way back into the city following the exuberant news of recent Allied victories in France and Belgium. Newspapers had announced the end of the V-1 threat, proclaiming "The Battle of London is over." What the public didn't know was that the supply situation was about to slow the advance for a much longer period than even Allied generals expected. Moreover, the retreating German armies had completely destroyed the French and Belgian railway systems, which meant most of the supplies were still being transported by truck, and in much smaller quantities than were needed. The natural barriers of the terrain, along with man-made obstacles such as the Siegfried Line, were also inhibiting the advance. It would take weeks before enough food, ammunition, and other vital supplies could be accumulated on the front lines. But by that time, summer weather would be over; inclement weather could also cause many problems for the Allies.[7]

Along the coast of England, British radar stations attempted to track the flight of incoming V-2s in order to find out just where the rockets were coming from. It was thought that radar might disclose the launching points. Just after liftoff, the rockets could be seen as a momentary blip on radar screens. Sound-ranging and flash-spotting equipment were also employed in Belgium, but in the end, these efforts produced too little information to discern the firing locations. But on many occasions, the Dutch resistance was able to get maps and messages to the British detailing the locations and operations of the German rocket troops. Rising with a tail of flame and a deafening roar, a V-2 could be seen and heard by Dutch civilians for miles around. With the help of these reports, some fighter-bomber attacks were mounted, but these met with little success during the early part of the rocket campaign. At first Allied commanders were unaware of the rocket's mobility and continued to envision some type of fixed launching complex. It took British Intelligence quite a long time to discover just how extremely mobile the rocket was. Not only was the rocket impossible to stop after firing—it was also going to be extremely difficult, if not impossible, to stop on the ground.

Following disagreements with General Metz, Kammler assumed total control of the rocket batteries as divisional commander. On September 9, acting on orders from Himmler, General Kammler ordered Battery 444 to cease firing on Paris and transferred them to Group North. They were ordered to travel to the Dutch coast and set up operations on the Walcheren peninsula near the Scheldt estuary across from the Belgian border. German military commanders were keeping a close watch on the situation at the front lines. American, British, and Canadian troops had already moved into Belgium and were at the German frontier. Battery commanders were advised to be ready to pull out of their operational areas on short notice. General Kammler was fearful that the rocket troops might be overrun, and the last thing he wanted was for his new weapon to be neutralized before the rocket campaign could get rolling. In The Hague, Battalion 485 received orders to continue firing at London, but because of the Allied advance, they were on alert for a possible withdrawal to the east.[8]

Kammler's staff, which was thrown together at the last minute, was unable to cope with all questions arising after the start of the rocket offensive. Because of his military command inexperience, he made many elementary mistakes. For example, after the start of the offensive, Kammler could not get in touch with his troops. He knew nothing of their whereabouts, particularly Battalion 485 in Holland. Some of the most essential equipment and spare parts had been left behind. On September 5, the day designated for the offensive to begin, the troops discovered that no warheads had been shipped to the operational areas, which postponed the opening of the offensive for several days. The battalions went into action with specialized tankers, trailers, and trucks to carry the rocket fuels, but the motorized columns had no tankers along to carry regular gasoline for their vehicles. When told of the problem, Kammler suggested the troops should just fill up the emptied alcohol tankers with gasoline. Immediately, the coating inside of the tanks—specially made for the storage of alcohol—was completely ruined. The alcohol tankers, which were in short supply and needed at the front, had to be sent back to Germany for repairs.[9]

Even though the rocket attacks had begun, the concerns of the launching crews were far from over. In the first ten days of the V-2 campaign, only two dozen rockets had been launched toward London. Fuel and supplies, especially the highly explosive liquid oxygen, were being brought in from Germany in frustratingly small amounts. The rockets were another problem. By the time they reached their launching sites, more than half were not fit for firing. Of those that were launched, only about 60 percent found their way to London. The familiar airburst problem, along with in-flight mechanical problems, caused many rockets to fall in the North Sea or in sparsely populated areas. Daily technical reports coming to headquarters reported a variety of problems such as oil in the liquid oxygen. This occurred when the liquid oxygen had been stored too long in the transport bowser. The oil residue caused explosions upon ignition of the combustion unit. A thorough cleaning of the bowsers cleared up this problem and stopped the explosions, but damage to materials and to the launch sites caused further delays. Long storage was also causing problems with the inner workings of many rockets. The delicate servo mechanisms were especially vulnerable to the effects of rain and humidity. At this time, the rockets were shipped by train from the Mittelwerk factory at Nordhausen to the area of Koblenz, where they were stockpiled in depots along the northwest German border. Once at the border, LXV Corps transported them to the main storage facility where the 511 Field Workshop Company unloaded and stored the rockets until they were required. There they remained for weeks at a time. When they were finally brought over to the launching crews, many of the V-2s were in poor condition, with vital components and electrical systems corroded. As a result of these conditions, the rate of fire was well below Hitler's imagined onslaught.

After being transferred to Group North, Battery 444 began its move from Belgium to Walcheren on September 10, arriving late in the evening on September 15. Along the way, as the rockets passed through the town of Serooskerke, a young Dutch girl, peeking through the drawn window curtains of her home, secretly snapped three rolls of film showing the tarpaulin-covered missiles resting on their long Meillerwagen trailers. The film rolls were later passed on to London, giving British officials their first up-close look at the V-2. At Walcheren, the Battery 444 firing crews quickly set up operations on the scenic grounds of the Vrederust estate, known today as Welgelegen. On the days of September 16–18, 1944, they launched six rockets at England, but only three reached their target.

On September 9, 1944, the staff at British Field Marshal Bernard Montgomery's headquarters received an urgent message from the Vice Chief of the Imperial General Staff: "Two rockets, so called V-2, landed in England yesterday. Please report most urgently by what approximate date you consider you can rope off the coastal area contained by Antwerp-Utrecht-Rotterdam. When this area is in our hands the threat from this weapon will probably have dispersed."[10]

On September 10, 1944, Supreme Allied Commander General Eisenhower flew to Brussels for a meeting with Field Marshal Montgomery. It was a trip he was not looking forward to. He did not think highly of the egotistic British field marshal who had been openly critical of Eisenhower's "broad front" strategy. Even while the Allied advance was grinding to a halt, Montgomery had been proposing what he called his "single thrust plan," an idea for a powerful thrust into Holland that would trap German forces in the Ruhr and allow the Allies to drive straight to Berlin, ending the war before Christmas. Eisenhower had heard this proposal before; he had rejected it several times already. Eisenhower believed the best way to get the Allied advance underway again was to open the Belgian port of Antwerp. If the estuary could be cleared, and the remaining resistance eliminated, the deep-water harbor would allow the necessary supplies to quickly reach the front lines. Eisenhower believed enormous stockpiles would need to be built up prior to the commencement of a long drive across the German heartland.

Eisenhower's meeting with Montgomery started off just as he had expected; the Field Marshal began lambasting the current Allied strategy and came up with all the same criticisms and optimistic proposals. Montgomery insisted if he received all the men and equipment he requested, he could capture a bridge over the Rhine and be in Berlin in less than three months. However, this time the British field marshal offered a different argument. Montgomery pointed out that the German rocket campaign had begun and that British Intelligence knew the rockets were being launched from somewhere on the Dutch coast. If the plan worked, not only would it be a catalyst into Germany but it would also counteract the V-2 threat. Holland had been under German occupation for four years, and the British commander believed that the German forces there were weak. If airborne units could land and hold key bridges, he could send a heavy armored force racing through Holland all the way to the IJsselmeer. This time Eisenhower agreed to the plan. Even though he was doubtful as to whether the operation would swiftly facilitate a passageway into Germany, the prospect of capturing a bridge over the Rhine, while at the same time reducing the rocket threat, must have appealed to him.

Code-named Operation Market Garden, it would be the largest airborne drop in military history. Three Allied divisions would be involved. In the "Market" portion of the plan, the U.S. 101st Airborne Division Airborne would drop near Eindhoven and secure the canal crossings at Veghel. The U.S. 82nd Airborne Division would capture the bridges over the Maas River at Grave and the Waal River at Nijmegen. Sixty miles behind German lines, the British First Airborne, then later the Polish First Airborne Brigade, would be dropped on the bridge over the Rhine River at Arnhem. In the "Garden" phase, British XXX Corps would dash up these Allied-held river crossings to relieve the First Airborne at Arnhem.

Montgomery's original plan called for the capture of a bridge across the Rhine at Wesel, a move that might have cut off Division z.V. headquarters near Nijmegen and allowed for the capture of SS General Kammler himself. But following the communiqué from London concerning the first V-2 attacks, Montgomery altered his plan in an attempt to cut off the V-2 supply routes to the Dutch coast. Subsequent to their retreat from France and Belgium, German forces in the west were beginning to stabilize. German paratroopers and SS Panzer units had moved into Arnhem, and while British intelligence was aware of them, for some reason, their presence was deemed insignificant by British planners.

Just after noon on September 17, 1944, 12 Spitfires of No. 229 Squadron RAF took off from their base at Coltishall. Their mission was to scan a stretch of the Dutch coast looking for evidence of rocket activity. While patrolling at 12,000 feet over North Holland, they witnessed a V-2 rising in the distance at terrific speed. The rocket impacted several minutes later in greater London at Coulsdon. The Spitfires were too far away to discern the exact location from which the rocket had been fired. They could only report the general location, which was near the coast, possibly The Hague. Battery 444 launched a rocket from Walcheren that came down at Adelaide Road, Brockley, Lewisham, killing 14 people and injuring another 41. Later that evening, more fighter bombers, acting on a tip from the Dutch underground, attacked the surrounding area near Beukenhorst between Raaphorstlaan and Eikenhorstlaan. These heavily wooded areas southeast of Wassenaar were suspected as firing locations but in fact were not.[11]

On the morning of September 17, 1944, the airborne landings began. The Dutch population, convinced they were about to be liberated, watched the armada of aircraft from their rooftops. Even the German troops were in awe of the force that was descending upon them. But as XXX Corps began to advance up a single road, the assault was stalled almost as soon as it began. German defenders immediately poured fire down on the tanks and vehicles. The airborne forces were able to accomplish their goals, except for the 82nd Airborne, which had to scramble to build a temporary bridge for XXX Corps. The British First Airborne had dropped five miles from their target and could only take the north side of the bridge at Arnhem. Worse, they had dropped in the middle of a Panzer division and were facing German attacks on their flank and continuing armored assaults from the south side of the bridge. They were cut off, and because of inadequate radio transmitters, the division commander was out of touch with his men for 36 critical hours.

First Airborne held on to their position with the hope that XXX Corps would eventually show up, but after nine days, no relief had arrived. Casualties were mounting, and it became clear that XXX Corps would never make it to Arnhem. On September 26, Montgomery ordered the First Airborne to break out of Arnhem and rejoin the Allied lines to the south. Out of 10,000 men dropped into Arnhem, only 2,300 came out; 1,400 were dead, and over 6,000 were prisoners of war. Operation Market Garden had failed. Montgomery's planning had not take into account any of the lessons learned in Allied operations of the recent past. For such a large operation, very little time was given to actual planning. Bad weather, bad intelligence, and stiff German resistance all contributed to the failure.

On September 19, 1944, at the beginning of the Allied airborne landings, SS General Kammler ordered the evacuation of all rocket troops from The Hague and Walcheren, for fear they might be cut off. The inhabitants of Wassenaar were able return to their homes after Battalion 485 withdrew from the area under the cover of darkness. The vehicles of the first battery traveled north and arrived at Overveen near Haarlem and then retreated all the way into Germany in Burgsteinfurt, where they were joined later by the second battery. The first battery of 485 set up operations west of the small town of Legden with two firing sites at Beikelort where they launched a total of 21 rockets from September 21 to October 8 against continental targets such as Louvain, Tournai, Maastricht, and Liège. At the beginning of Market Garden, American forces almost nabbed General Kammler at Berg en Dal, so he also moved his headquarters to the German town of Darfeld in Burgsteinfurt for a short time; but after moving again to Ludenscheid on September 21, he soon established a permanent headquarters in Germany, east of Dortmund at Suttrop bei Warstein on October 3.[12]

It is often reported that Kammler's headquarters was located for a time in the Dutch town of Haaksbergen. In fact, there was no German headquarters of any kind at Haaksbergen. It may have been confusion between the names Haaksbergen and Schaarsbergen. Schaarsbergen was about 20 kilometers from Apeldoorn, and many German barracks were concentrated in this area. Dornberger and Kammler reportedly met on several occasions at a location near Apeldoorn.[13]

In early September of 1944, after their instruction at Köslin, SS Werfer Battery 500 moved into Poland for live firing exercises. The SS rocket troops were brought to the V-2 test range at Heidekraut, where they rapidly received their final training. Crew members were hastily rehearsed in their individual duties. Finally, the test missiles were moved to a remote launching site, fueled, tested, and fired. Beginning on September 13, each platoon launched several missiles, but lack of motorized transport and ground support equipment hampered portions of the training. All new equipment delivered had to be allotted to the operational units on the battlefront. When the exercises concluded on September 19, the V-2 campaign was already underway, and the whole battery, vehicles included, was loaded onto a freight train that departed toward

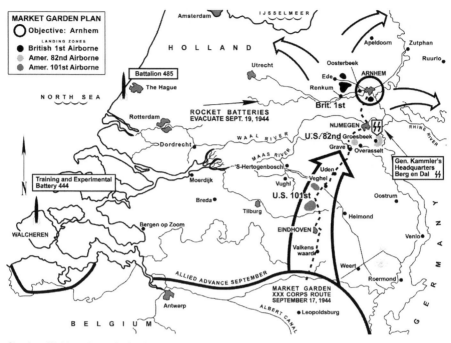

Situation of V-2 batteries on the Dutch coast at the beginning of the Market Garden offensive, September 17, 1945.

the west. The train arrived in Münster, where everything was off-loaded. From there, the SS 500 drove further to Burgsteinfurt. They moved into a prepared launching site code-named Schandfleck among the tall oak trees just off the main road to Schöppingen near the German town of Heek. The battery fired its first operational V-2 on October 19, but the rocket crashed in a meadow only a few miles from the firing point. During the next two weeks, the SS 500 launched approximately 30 more V-2s, all successfully.[14]

The first battery of Battalion 485 had already been operating in Burgsteinfurt since September 21 and was reported to have launched ten times against Liège over the period of October 4 to October 12. In total there were 27 reported impacts in the surroundings of Liège.

After the third battery of Battalion 485 completed their training at Heidekraut, they were sent into action in mid-October 1944. They joined SS 500 and the first battery of Battalion 485 in the Burgsteinfurt area, driving to a firing location west of the German town of Legden at Beikelort. The battery at that time consisted of only the first and second platoons; the third platoon had been sent to The Hague, which was very unusual, because normally individual platoons were not separated from their batteries. From Beikelort the two platoons fired 35 rockets at Antwerp over the period of October 21 to November 5. The railway siding outside the town of Legden was the main supply point in the area, and on October 9 Allied fighter bombers attacked a V-2 transport train, which was burned out and destroyed. At the beginning of November, the two platoons moved to a new position just outside of Legden near the castle Schloss Egelborg. In late November, the second platoon relieved the third platoon in The Hague; but a few weeks later, all three platoons were once again together near Legden in Burgsteinfurt.[15]

On September 15, Group South began firing from positions in Germany near Euskirchen about 20 miles southwest of Cologne. Their targets included cities of varying importance: Arras, Amiens, Tourcoing, Brussels, Mons, Charleroi, Lille, Liège, Tournai, Diest, Hasselt, Maastricht, Cambrai, and Roubaix, all of which were soon to be in Allied hands. Battalion 836, under the command of Major Weber, continued in the Euskirchen area at sites Rheinbach and Kottenforst until the last week in September, when it moved across the Rhine River to the Westerwald area north of Montabaur.[16]

From September 21 until October 3, there were four launching positions in the area of Westerwald (east of the Rhine River near Koblenz)—two near the town of Roßbach and two near the town of Helferskirchen. The command post of the third battery of Battalion 836 was situated in the Roßbacher Forest (District of Herrlichkeit), while the soldiers were billeted at Roßbach and Mündersbach. At Roßbach the launch sites were just off the main road with new paths created in half-circle tracks of 50 yards into the beech trees and then back onto the main road, which made it very easy for the long Meillerwagens delivering the rockets to maneuver in the wooded area. There were two firing pads there consisting of leveled ground covered in gravel. The second battery of Battalion 836, along with the headquarters battery, the technical battery, the flak train "Wojahn" (armed with antiaircraft guns), and the security platoon Sendezug/Funkhorchkompanie 725 (a unit for observation of trajectories and jamming signal) were established at Helferskirchen where the firing location was positioned about three-quarters of a mile south of the town. Under the standard of the battalion's symbolic unit crest, which featured the figure of a witch riding naked on a broomstick, the third battery of 836 launched two rockets the next day toward Liège. These were the first V-2 rockets fired from the Westerwald. Only one was successful. The second rocket launched from Roßbach was a failure. The troops heard a huge thud at ignition; nonetheless, the rocket lifted normally into the sky only to explode after 40 seconds of thrust.[17]

On September 27, 1944, enemy aircraft flew over the V-2 positions in the Westerwald area. Twelve Allied fighters bombed and strafed the railway stations at Hattert and Hachenburg and also the flak emplacements of the third firing battery, but no significant losses were reported. The battalion headquarters was soon established in Hachenburg, while the technical troops were stationed in the forest north of the road from Steinen to Dreifelden. From the railway station at Selters, the rockets were moved through Herschbach into the forest near Marienberg, where the Technical Troop checked and serviced them. After the warhead was mounted, each rocket was towed by the Meillerwagens over forested roads through the town of Roßbach and the Hachenburger Way to the firing sites.

The second battery of 836 fired seven V-2s in a 27-hour period, on September 27–28, but the site was abandoned the next day, and they were ordered to new emplacements across the Rhine to Saarland near Merzig (not far from the old training ground

V-2 Firings September 8-17, 1944

GROUP SOUTH

Firing Units	Targets	Approx. No. of V-2s Fired	Details
Battery 444	Paris	2	Began operations from the Ardennes Forest in Belgium on September 8, 1944, after two unsuccessful launch attempts the previous day. Moved to Dutch coast on September 10.
Second Battery Batt. 836	Lille	4	Began firing on September 14 near
Second Battery Batt. 836	Mons	3	Euskirchen.
Third Battery Batt. 836	Lille	3	
Third Battery Batt. 836	Mons	1	

GROUP NORTH

Firing Units	Targets	Approx. No. of V-2s Fired	Details
First Battery Batt. 485	London	12	First rockets fired from sites
First Battery Batt. 485	Liège	1	in the neighborhood of Wassenaar.
Second Battery Batt. 485	London	12	After ten days of operations the batteries withdrew to Burgsteinfurt, Germany as a result of the Allied airborne landings near Arnhem.
Battery 444	London	6	Redeployed by SS General Kammler to Walcheren on Dutch coast where crews launched September 16-18.

V-2 Firings September 18-30, 1944

GROUP SOUTH

Firing Units	Targets	Approx. No. of V-2s Fired	Details
Second Battery Batt. 836	Lille	7	Firings from area near Euskirchen until end
	Arras	5	of September, then moved across the Rhine
	Cambrai	5	River to area near Montabaur in the
	St. Quentin	1	Westerwald region of Germany. On
	Tourcoing	5	September 29 moved again to Merzig and
	Liège	3	began firing in October.
	Hasselt	1	
	Maastricht	2	
Third Battery of Batt. 836	Lille	3	Firings from area near Euskirchen until end
	Liège	5	of September, then moved across the Rhine
	Arras	1	River to area near Montabaur in the
	Maastricht	2	Westerwald region of Germany. Followed
	Tourcoing	1	the second battery to Merzig on October 6.
	Tournai	2	
	Hasselt	3	

GROUP NORTH

Firing Units	Targets	Approx. No. of V-2s Fired	Details
Battery 444	Norwich	43	Evacuated to area of Rijs in Gaasterland
	Ipwich	1	launching near the waters if the IJsselmeer and began firing on September 25.
First Battery Batt. 485	Tournai	1	Firings from sites in Burgsteinfurt starting on
	Hasselt	2	September 19.
	Maastricht	8	
	Liège	4	
Second Battery Batt. 485	Tournai	6	Launched from Burgsteinfurt up to
	Hasselt	4	September 30 and then moved to
	Maastricht	4	The Hague and began firing on October 4.
	Liège	8	

ground at Baumholder) to begin firing on Paris. General Kammler had ordered that all elements of Group South should concentrate their fire on the French capital. On October 3, the third battery, stationed at Roßbach, was ordered to join them; and by October 6, both batteries were in the Merzig area.[18]

In the summer of 1944, as the Red Army was closing in on the Heidelager test range, R. V. Jones suggested to Winston Churchill that a technical team be sent to Poland to investigate the facility. Churchill agreed and immediately contacted Marshall Stalin to request access to the area at Blizna. Jones had assumed that the Air Ministry would be responsible for the mission; however, ultimately it was the Crossbow Committee that was charged with the Big Ben mission to the Soviet Union. A team of technical advisors led by Colonel T. R. B. Sanders, an engineer with a practical knowledge of ballistics known for his discretion and likeability, was to lead the Allied team. The relationship between Britain and Russia during the course of the war was marked by significant reservations. The appointment of a team leader having both substantial tact and diplomacy was important. Considerable negotiating skills, along with the ability to foster favorable relations, would be needed to secure the assistance of the Russians.

After permission was finally granted by the Russians, the party spent approximately two weeks in Teheran while waiting to get authorization from the Russians to fly to Moscow. Once in Moscow, another two weeks elapsed before the team actually reached Blizna on September 3. The party consisted of a mix of British and American personnel, who got on well together as well as with the Russians. Upon their arrival at the Heidelager camp, heavy fighting could be heard in the distance, as the front lines were only five miles away. The team quickly set about taking measurements, inspecting craters, collecting useful bits of rocket scraps, and interrogating Polish citizens. The firing platforms were the most remarkable discovery at Heidelager. Colonel Sanders was surprised to find out an area of only 20 square feet could be used to launch the V-2. The team managed to find a combustion chamber, a segment of the turbine casing, one of the fins, the framework of a radio compartment, and some smaller items, in all more than a ton of parts. Conversely, no evidence of railway wagons of any sort or storage tanks could be found. Looking for bits of rockets, some of the men also traveled to the target area to examine the impact sites. It was found that the Germans had been very thorough in collecting fragments, however. Of the many impact craters inspected, not a single large piece of scrap was found in the target areas. All of the collected material was crated up in Blizna and trucked to Moscow, where it would be shipped to Britain.

Colonel Sanders and his colleagues left Blizna on September 20, 1944, arrived in Moscow two days later, and flew back to London. As a result of the mission, it was found that there could be no countermeasures for the V-2. On the other hand, the rocket was found to be much smaller than anticipated, with a one-ton warhead, rather than the ten tons that some had predicted. Weeks later, a message from Moscow arrived in London with information that the rocket parts from Blizna had been "temporarily lost." The crates were eventually sent on; however, when they were opened, authorities found not rocket parts but rusting scraps of motor cars. The Russians, after learning of the German long-range rocket program, had developed their own agenda.[19]

After spending only two days at Walcheren, Battery 444 was ordered to travel north to Gaasterland in southwest Friesland, where it could continue operations against England. The battery traveled under the cover of darkness, as it was very risky to be on the roads during daylight hours because of Allied air superiority. After arriving in Friesland, Battery 444 set up operations in a small forested area called Rijs, south of the city of Balk. About 30 to 35 German officers of the battery were billeted in the nearby Hotel Jans. The Rijsterbos (Rijster Forest) was just off the waters of the IJsselmeer (Zuider Zee), a huge shallow lake in the center of Holland. Moving by train, the rockets left the Assen railway station bound for the station near Heerenveen at

Sneek. At Heereveen they were placed on Vidalwagen road transport trailers of the supply troops and then towed by truck via the towns of Hommerts, Woudsend, and Harich, then into Balk. Dutch residents witnessed many vehicles and rockets parked beside City Hall in Balk. The rockets had to pass over a small bridge and make a difficult turn via the roads Van Swinderenstraat and Houtdijk. The Germans cleared trees away for maneuvering the trailers over the bridge that lead to Kippenburg. Even today, one can still see the scratches on the bridge where the V-2 trailers clipped the railing upon making the turn. At Kippenburg the rockets were prepared with their warheads and transferred to the Meillerwagen erector trailer. Behind the large estate house at Kippenburg, the propellant and warheads were stockpiled. The rockets were then moved via Rijsterdyk and Murnserleane a few kilometers southwest to the launching sites. Strung over the dark, unpaved, forested lanes of Murnserleane and Middenleane were large camouflage nets suspended high in the trees for further concealment from Allied aircraft.[20]

With the V-2 having a maximum range of approximately 200 to 230 miles, it was not possible to target London from the location at Rijs. Instead, Battery 444 turned its attention to East Anglia and the territory surrounding Norwich in eastern England. Kammler was determined to continue the strikes on the British public from wherever possible, even it if meant targeting lesser cities.

On September 25, at 18:05 hours, after the trees and shrubs were sprayed with water to lower the fire danger, Battery 444 launched its first rocket toward northern England. Approximately five minutes later, the rocket hit a farm field at Hoxne in Suffolk, inflicting only minor damage to a few buildings nearby.

That same day, the rocket troops encountered their first misfire. A rocket had to be drained of its remaining fuel after the engine failed to generate full thrust. The ignition cable was burnt as the engine continued to fire while not leaving the launch table. Upon inspection, it was discovered that the rudders and tail section had been severely scorched, so the rocket was sent back for refurbishing. Closer investigation of other rockets from the Mittelwerk had revealed many additional problems. Bad welds, missing parts, short-circuited electrical connections from inferior soldering—these were just some of the mechanical errors discovered. Not only did the crews face difficulties from the quality of the rockets, there also existed an acute shortage of liquid oxygen. German production had only reached a level of about 200 cubic meters per day, which is only enough to launch 24 rockets. The logistical problems of firing batteries on the move and V-2 units spread out from northern Holland to western Germany did not help matters.

Late in the afternoon on September 26, a loud double boom was heard near the English village of Ranworth. The rocket plowed into a field about eight miles outside of town. The sound of the explosion was followed by another loud sonic boom and then the whine of rushing air. Windows of cottages were shattered within a half-mile radius of the blast. Officials in Britain quickly knew that the V-2 campaign had come to East Anglia. There had been a reduction in the frequency of V-2 attacks since the beginning

V-2

Europe during the V-2 campaign of late 1944 and early 1945.

SWEDEN

Baltic Sea

Danzig

Peenemünde

Köslin

Heidekraut

Rostock

Berlin   Posen

Warsaw

A Z I   G E R M A N Y   GENERAL
                                   GOVERNMENT
Mittelwerk   OF POLAND
Nordhausen   Leipzig

                                   Blizna
                                   (Heiderlager)

Lehesten
                                   Krakow

              Prague

      BOHEMIA AND MORAVIA

                                   S L O V A K I A

Munich   Vienna

   Salzburg   Wiener
                Neustadt   Budapest
Oberammergau   Ebensee

                                   H U N G A R Y

of the Market Garden offensive, but still no word concerning the nature of this new German weapon had passed from British authorities to the populace. These mysterious bangs were new to the citizens of Norwich. Even some of the nearby military establishments were unfamiliar with the new threat and recorded these first impacts as aircraft crash sites.[21]

At Rijs, the Dutch citizens were unsure of just what was going on near their homes. They only knew that it was a dangerous operation. The entry lanes to the forest were strangely blinded with canvas. They could hear on German radio the propagandists heralding the new Wunderwaffen (wonder weapons) but were unsure what this meant. Weeks later the BBC reported that it was the V-2 rocket falling on England. It was forbidden to come close to the launching sites, and very few people risked being caught near the area. Not only was there the danger of the German guards, there also was the peril of failed rockets crashing in the immediate area.

On the afternoon of September 30, a V-2 was launched from Rijs. It rose to a height of 600 feet before an explosion in the rocket's tail brought it crashing to earth about 20 yards from the firing table. The alcohol and liquid oxygen tanks exploded upon impact, injuring some of the firing crew. The warhead sizzled in the burning fuel and exploded approximately 45 minutes later, digging a huge crater. This failed rocket had ironically destroyed a small shrine in the forest called Vredestempeltje (the little peace temple). Despite the mishap, a new launch site was quickly established a few hundred yards away.[22]

Several weeks later, Wieger Jurjen Draayer, a local farmer, was riding his bicycle along the lanes just beyond the Rijs Forest near Bakhuizen. As there had been strange noises and unknown things seen in the sky for the past several weeks, Wieger was anxious to get home. In the distance he suddenly heard a thump followed by a tremendous roar. The bicycle he was riding came to a stop, and he let it fall to the ground. Racing to a nearby ditch, he peered out to witness a huge steeple-shaped object trailing a tail of fire rising from the forest ahead of him. The object was arcing above him when something went wrong. The noise from the projectile ceased, followed by a whistling as it fell from the sky. There was a tremendous explosion some 70 yards from where Wieger hugged the side of the ditch. Quickly, Wieger got on his bicycle again and started peddling as fast as he could. Tiny bits of material were floating down all around him, almost like snow. He noticed three dead cows in the field near the forest. The explosion left a crater some 20 feet deep and 30 feet wide. As he approached an intersection, a group of German soldiers called for him to stop. The soldiers were surprised to see the farmer riding so close to the V-2 launching area. They asked if he was injured and told him this was a restricted area and to stay away in the future. The dazed and confused Wieger hurried to his home.[23]

For the Dutch residents of the surrounding countryside, it was a very nervous time. Every day they could hear the thunderous noise of the V-2 launches and lived in fear that something might go wrong. The farmers soon knew if the rocket did not rise vertically, anything could happen. Failed rockets would fall in the immediate area, some-

Sequence of photos showing a ragged-camouflage V-2 rising from the forest floor. (*NARA*)

times near the residents' homes. Other V-2s encountered problems at higher altitudes, and the farmers watched them plunge into the waters of the IJsselmeer just off shore.

For the soldiers of Battery 444, the stress of the launches was just as great. Many of them would rather have been occupied with some less hazardous job. However, there was plenty of Dutch gin to help them ease their tensions. British fighter planes searched the area several times; however, the ability of Battery 444 crews to launch and retreat quickly made it difficult to spot anything from the air. The Rijs Forest V-2 sites, with very tall trees, provided excellent camouflage; but there was always the possibility of an air attack, and the rocket troops were very wary of this.

On October 3, marking the second anniversary of the first successful A-4 launched from Peenemünde, the rocket troops at Rijs fired six missiles toward the Norfolk countryside. Throughout the day, thunderous detonations reverberated at regular intervals. From their homes, the people of Norwich could see huge columns of black smoke in the distance rising high into the air. The strikes were gradually coming closer to the populated sections of the county. Late that evening, an explosion rocked the Hellesdon area. An estimated 400 houses within a two-mile radius were damaged in some manner. The following day British authorities recovered the remains of a V-2, which broke up in the air before impact near Spixworth. The engine and various important parts were sent to Air Institute at Farnborough for analysis.[24]

The last rocket to fly toward England from Rijs was launched on the morning of Thursday, October 12. It fell innocuously in the open near Ingworth without much commotion, demonstrating the folly of targeting anything less than a large urban city with the V-2. From September 25 to October 12, Battery 444 launched approximately 43 rockets toward East Anglia. Without heavily populated English targets within range, the results were not satisfactory. British casualties from V-2 attacks in East Anglia ended up relatively light. Only one person had been killed as a result of the attacks, and less than 50 people were wounded. The damage in Suffolk and Norfolk counties was limited to only a modest amount of houses, barns, farms, and schools. Many V-2s struck empty fields and even the North Sea.

The campaign against East Anglia ended on October 13 after new orders were received to begin targeting the port of Antwerp. Most Battery 444's initial shots toward Antwerp missed their mark, falling short in and around the suburbs of the port city. However, on October 16, 1944, they scored a direct hit, when a V-2 slammed into dock number 201 in the harbork, completely demolishing it.

In October, RAF Fighter Command began a new operation designed to impede the German rocket crews in Holland. Big Ben patrols, or anti-V-2 missions, were mounted by several RAF squadrons, which used armed reconnaissance and dive-bombing sorties to attack the rocket areas on the Dutch coast. Flying out of Coltishall, No. 602 Squadron RAF was brought in to patrol for V-2s on October 10. They joined in with No. 229 Squadron RAF, which had already been in operation against the rocket sites since September. Both the RAF and the U.S. Army Air Force did their best to locate the missiles, and on many occasions sent hundreds of fighters over Holland to strafe anything that looked like a target. However, because of the rocket's surreptitiousness and the heavily camouflaged firing positions, it proved almost impossible for pilots to locate the rocket batteries on the ground. Even so, the fighter-bomber sweeps shot up a lot of vehicles and railway cars and were partially responsible for shortages of liquid oxygen and other supplies at the V-2 launch sites.[25]

After three weeks, Battery 444 disappeared from Gaasterland just as quickly as it had arrived. The last rocket fired from Rijs headed for the port of Antwerp on the morning of October 20. Suddenly the Germans packed and moved south that same day. SS General Kammler had ordered the unit back to The Hague following the failure of Market Garden.

Ever since the first rocket was fired from Rijs, British radar momentarily tracked the incoming missiles. In addition, Allied pilots reported sightings of contrails from ascending rockets near Gaasterland. However, these only gave an approximate location of the firing positions. After several weeks, an RAF reconnaissance aircraft brought back a photograph showing clear evidence of activity in the forest. On October 21, a flight of seven Tempest fighter bombers of the No. 274 Squadron RAF flew near the Rijs Forest and finally located the launching sites. They flew by heading east, just north of the forest, and after forming up in a line, the seven aircraft turned back to attack the area. Not only did the aircraft drop bombs in the forest, they also shot up the surrounding houses and buildings. Luckily, farm animals were the only victims of this attack, although some civilians narrowly escaped being hit. But the RAF found the launching sites only a few hours after the last Battery 444 vehicles exited the area. The British were unaware that the rocket units were gone, and the bombers returned each of the next few days to attack the forest. By this time, the firing platoons of Battery 444 were arriving in The Hague to join Battalion 485 for operations against London.

The people around Rijs were relieved to see the German V-2 menace departed. They returned to their everyday life, as it was in wartime, without the threat of exploding missiles on their homes. Actually, they were very lucky. There had been no Dutch civilian casualties. If not for the light population of the area and the fact that the missiles were traveling over the IJsselmeer after launching, the casualties may have been severe. Moreover, the relatively short three weeks of operations meant there was very limited damage. Later, on clear winter days, they could see the V-2s rising from the Eelerberg over 100 kilometers away to the south. They could easily imagine the terror felt by their neighbors there, who must be enduring the same nightmare they experienced only a few months before.

Now that Montgomery's offensive had been defeated, the V-2 batteries which had retreated to Germany began returning to The Hague. They selected the large, adjoining, open areas of Bloemendaal and Ockenburgh, far removed from the built-up city center, for their firing sites. On October 3, the first Meillerwagen of the second battery of Battalion 485 drove into Ockenburgh at 9:00 in the morning. Later that evening, they launched a V-2 at 11:00 PM, followed by another 45 minutes later, which exploded shortly after liftoff, lighting up the whole city. The V-2 had returned. On October 7, V-2s began being launched from the Bloemendaal site. Because of the launch site locations, the RAF decided they could bomb them without risking too many civilian casualties. After several days of bad weather, six aircraft dropped their bombs on Ockenburgh and Bloemendaal in the early morning of October 18. At Bloemendaal the bombs damaged a rocket resting on a Meillerwagen that was unconcealed, out in the open.

In early October, German commanders first considered the possibility of firing V-weapons against Antwerp. The objective would be to smash the harbor installations so effectively that they would be useless to the Allies even after the approaches to the harbor had been cleared of German resistance. On October 12, 1944, the Führer ordered all V-2 fire be directed at London and Antwerp exclusively. Attacks on all other targets would stop.

6

Twilight of the Gods

In the autumn of 1944, Rita Winter was a young Dutch girl living with her mother and younger brother on Damastraat, near the suburb of Rijswijk, in The Hague. One day Rita's little brother, Frans, came home telling a wild story to his mother about how they must leave their home immediately. When questioned further, the boy said he met some German soldiers while playing that afternoon. One of the soldiers had told him that a "special new weapon" would be fired that evening around 8 o'clock, and they were unsure if the ground was firm enough to support the weapon. The soldiers told Frans he should leave his home because there was a chance the weapon might topple over and explode, destroying nearby houses. Rita remembers that her mother laughed very hard, thinking the young boy was fabricating the story. "Yes, sure they told you that!"

By October 21, 1944, Battery 444 had joined the second battery of Battalion 485 for operations in The Hague. They returned to the grounds of the Duindigt estate at Wassenaar where V-2s had been launched before to begin firing at London on October 22. On October 23, Battery 444 established a new firing point near Rita Winter's home, just past the gardens and the villas, on the grounds of the Rijswijkse Bos (Rijswijk Forest). A secluded clearing already existed in the wood where a monument, built in 1792, stood on the previous location of the castle of the Prince of Orange. It was here in 1697 that the peace treaty to end the War of the Grand Alliance was signed between England, France, Spain, Germany, and Holland. The new V-2 launch site was only a few yards from the obelisk to peace.

Rita remembers that her mother paid no more attention to her younger brother's "exaggerations" and went about her daily chores that evening. At exactly 8 o'clock, they were startled by the thump and roar of the V-2's engine coming to life. Rita thought her lungs would blow apart. Her mother ran to the window in time to see the rocket rise over their home, the red glare from the engine exhaust lighting up the streets as if it were daytime. The noise from the engine stopped and few seconds later the rocket went down with a flash on the west dune at Scheveningen. Later that night, several more rockets were fired.[1]

During the first three weeks of October, usually one or two rockets were launched per day; but now with Battery 444 also firing from The Hague, five firing tables were in operation daily and no less than six to seven missiles were launched each day. Immediately after sundown, a stream of V-2s were being pulled down Vredeburgerweg in Rijswijk to the new launching site. On October 26, a new launch record was achieved by the two platoons of Battery 444. A total of nine rockets were fired this day from the sites at Wassenaar and Rijswijk, three within 25 minutes of each other.

From their homes, the citizens of The Hague witnessed many rockets soaring skyward. In the beginning, watching the missiles roar off their firing tables was awe-inspiring for Dutch onlookers. However, for the German crews, preparing a V-2 for firing meant hard work and tense nerves. The V-2 was a sophisticated weapon. Even the smallest error might cause it to malfunction and crash or blow up on the ground. Many of the launches in The Hague took place in territory called Das Sperrgebiet (restricted area). One of these was a three-kilometer strip of land along the coast where all residents had been forced out earlier because of construction undertaken for Hitler's Atlantic Wall. Security measures along this strip were very strict. The Germans wanted to prevent prying eyes from witnessing launch preparations, but civilians still managed to get close enough for a good look. Some young children were free to go in and out of the restricted areas, such as the Haagsche Bosch city park, to gather firewood for their families and could go into the Sperrgebiet without being bothered. Once inside the restricted zone, some of the children became acquainted with the German soldiers. They would strike up conversations with the uniformed men and were able to learn detailed information about the rockets. At times, the children secretly inched dangerously close to the rocket to observe an actual launch.[2]

On October 27, the results were not so good. Around 2 o'clock in the afternoon, from the firing site at Beukenhorst, Battery 444 launched a rocket that climbed steadily to a height of about 280 feet before the engine abruptly shut down. The rocket fell back onto the launch site, destroying a portion of the crew's equipment. Two hours later, as young Rita Winter walked with her two little nieces down a street near her home, she heard the start of another V-2. This one, launched from the opposite end of Beukenhorst, rose straight up from Rijswijk, and then the engine began to stammer. The rocket oscillated back and forth before crashing to earth. The girls ducked into the window bay of a store front. The rocket came down about three-quarters of a mile away at Vredeburgerweg on a boy's school, killing five brothers and two others. Following the two explosions, the heavily damaged launching area was used no longer; in addition, the ammunition dump at Overvoorde was too close, so the launching sites at Rijswijk were abandoned that same day.[3]

In the following months, there were a lot of failures. In Rijswijk a rocket came down on the Roman Catholic Institute, killing 20 people. Another explosion in Voorburg at Koning Wilhelminastraat destroyed six houses, while another rocket came down on the railway station in Wassenaar. Dutch intelligence put the failure rate at about eight percent. Many of the rocket crew's reports coming to V-2 division headquarters sounded similar:

"First Battery Battalion 485, Serial No. 18374, September 22, 1944, 04:45 hours; Thrust terminated shortly after liftoff, unknown reasons. Device fell back on firing table and exploded. Cause not determined."

"Training and Experimental Battery 444, Serial No. 18979, September 30, 1944, 15:50 hours; After a series of standard checkups, the rocket was prepared and fired as usual. At an altitude of approximately 219 meters, there was an explosion in the midsection of the tail unit, which terminated the thrust. The rocket crashed about 20 meters from the firing table. The fuel tanks exploded, the warhead simmered in the burning alcohol for 45 minutes before detonating."[4]

"Training and Experimental Battery 444, Serial No. 18951, October 7, 1944; Rocket was returned to field store after preflight tests revealed problem with tail fin rudder. After inspection of the fin, the error was not clearly ascertained. Error is probably in trim motor. Since investigation requires the removal of the tail section, the device was returned."

During the month of October, 83 rockets had been launched, of which five were failures. This was better than usual; on average 8 to 12 percent of missile launches ended in failure. Some rockets blew up on their launch stands, killing and injuring crew members; some failed to ignite at all; others hung in the air for a moment, then crashed to earth and blew up or fell into the sea. Whenever Dutch civilians heard the roar of ignition, everyone would begin to count the seconds. After 30 seconds, they were safe; if the engine stopped after 30 seconds or more had gone by, the rocket would either crash into the North Sea or fall somewhere outside the city. However, if the roar stopped prior to 30 seconds, the missiles would fall back onto the neighborhoods of The Hague. Rockets were seen blasting wildly over the city, out of control, or shooting horizontally, only to crash a few miles from the firing site. Most exploded on impact; if the warhead did not go off, German specialists would try to defuse it. Many of the failed rounds fell on the residents of the city. The detonation of the warhead, along with the alcohol and liquid oxygen, destroyed hundreds of houses and caused many civilian casualties. The airburst problem had not been completely solved either.

After spending the summer stationed at Grossborn, the first battery of Battalion 836 moved to the Heidekraut test range on August 21, 1944, to finish their apprenticeship in firing operations. Following their V-2 shooting trials in Poland, the battery was shipped to the area of Westerwald, east of the Rhine River. On October 6, Fritz Siewczynski, who was a soldier attached to the first battery, arrived with his unit in the small town of Rennerod, where they unloaded all of the vehicles and launching equipment brought with them from Poland. After another short drive that same afternoon, they arrived at the small town of Beilstein (Greifenstein), a short distance from Herborn, seven miles east of Rennerod. It was the time of the fruit harvest, and at every corner local families displayed baskets of full ripe plums. Fritz and his buddies were allowed to eat as much plum cake as they wanted, even in those lean times of war. He had no idea as to the reason for their stay at Beilstein, but they would enjoy this beautiful rest from the war for the next 18 days. On October 24, 1944, they drove further south to the station at Herborn, where they loaded the vehicles again on rail cars of the National Railroad. The rail transport brought Fritz's unit to the area of their first military deployment. After arriving in Beuren, they took positions near Hermeskeil, about ten miles east of Trier.[5]

On October 14, the second and third batteries of 836 began firing upon Antwerp from positions near Merzig. Soon the rockets began arriving in the deployment zones. The trains brought the V-2s, riding on ten to 12 four-axle railcars, which were sometimes covered with gray, black, or white canopies or camouflaged with piles of hay, to the nearby off-loading points. The goal was to make the trains look like ordinary transportation of goods. On October 26–28, the second and third batteries also moved into the area of Hermeskeil and joined the first battery for operations against Antwerp.

Although the launching batteries in The Hague hit London 82 times in November, the crews of Battery 444 and Battalion 485 had no real idea where their missiles were striking. They could only set the gyro mechanisms and hope they were accurate. Several unsuccessful attempts were made to pinpoint the V-2 impacts by using radio waves and seismology; however, in the end, the best reports came from German spies, but even those reports were suspect, as British Intelligence had turned many of the spies around. Rockets frequently broke up in the upper atmosphere, high above the North Sea. On November 12, a rocket broke up in the air over London's Victoria Station. Astonished people saw a puff of smoke bloom in the sky, followed a few seconds later by a distant explosion and a hail of metal fragments.[6]

At the outset, when the first rockets rolled off the Mittelwerk assembly lines, they were plagued by bad welds, soldering problems, and faulty parts. Electrical components were installed, and final testing was performed at Degenkolb's DEMAG facility at Berlin-Falkensee. Later in 1944, these activities moved to Mittelwerk. Many of the slave laborers involved in electrical assembly and testing were required to put slips of paper bearing their unique identification numbers alongside the parts they had created or certified. Then, if problems were found with these during later inspections, the workers responsible would be punished. Still, minor forms of passive sabotage could be accomplished by the prisoners; for example, they accepted for assembly subcontractor parts that they knew did not meet specifications. There were instances of prisoner workers knowingly passing along electronic subassemblies that contained cold solder connections—ones that were likely to produce intermittent or no electrical contact at all, thus leading to failures. Other prisoners made only partial arc welds in obscured locations on the rocket such as inner welds on fins that would hopefully come apart later under launch stresses. But it was a dangerous enterprise; the penalty for sabotage was death.[7]

The SS guards often carried out individual or group hangings in the tunnels as a warning to the prisoners. The huge cranes in Hall 41 were used to hoist victims up by their necks and let them slowly strangle, in full view of the workers of each shift, who were specially called to witness these hangings. The dead were then left hanging there, about five feet off the floor, for a day or so, while the prisoners came and went beneath them. A permanent gallows was also erected in the roll call yard at Dora. The inhumanity would continue near Nordhausen at an even faster pace; on November 23, 1944, Hitler ordered the production of V-2s be raised to its maximum capacity of 900 units a month.[8]

During the first week of November, 12 V-2s hit London; during the second week, 35 V-2s came down; during the third week, there were 27 V-2 incidents. When the first few rockets landed on England in September, they had been little more than a nuisance. But in November several were hitting every day. By November 14, there were about four V-2 incidents per day. Under a new supply system called Warme Semmel, or hot cakes, rockets no longer sat in storage for weeks at a time before launching. It was hoped that by firing the rockets quickly—only a few days after coming off the assembly lines—the deterioration of vital parts would be prevented, thus reducing the number of failures. At Dornberger's headquarters in Schwedt, the transportation office planned the fastest routes for moving the rockets from the Mittelwerk directly to the front. In order to discover other potential problems, Dornberger's staff sent technical experts from Peenemünde to the launching areas to inspect and supervise the transport, handling, and delivery practices in the field. These technical crews maintained a supply of spare parts and introduced new components for the missiles. After a number of corrections, the overall amount of failures was reduced. Beginning in November, LXV Corps was only responsible for the V-1 operation. All former aspects of LXV Corps' involvement in V-2 operations had been swallowed up by Kammler. By November 20, 1944, about 210 rockets had reached England, with 95 hitting London. Four hundred fifty-six people had been killed in London alone, with hundreds more injured.[9]

Prior to the war, the pier at Scheveningen, which was the largest coastal resort in Holland, had delighted thousands of Dutch visitors each month. By 1910, Scheveningen had become one of the most fashionable resorts in Europe. It featured a broad seafront boulevard, prestigious hotels, and a concrete pier with tourist facilities. First mentioned in 1280 as the small fishing village of Sceveninghe, it was located about three and a half miles from the center of The Hague. Sadly, following the Nazi occupation of Holland in 1940, the seafront at Scheveningen had become entangled in Hitler's Atlantic Wall. Beginning in 1942, entire sections of the old town were leveled to make way for the construction of the new defenses. Even the pier was burned down in 1943. Gun batteries, tank walls, tank ditches, dragon's teeth, barbed wire, and fortified bunkers lined the beachfront, while German command posts occupied the remaining buildings. In November of 1944, an arm of the Dutch railway system ran all the way up to the resort, which made the area an attractive location for V-2 supply and logistics.

Officials from the V-2 training school at Köslin often visited The Hague to examine the actual situation concerning the transportation and handling of rockets and fuel

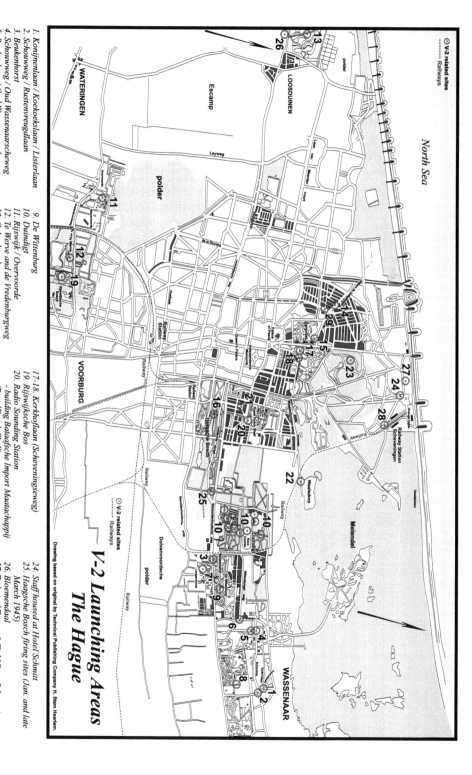

V-2 Launching Areas
The Hague

Drawing based on original by Technical Publishing Company H. Stam Haarlem.

1. Konijnenlaan / Koekoekslaan / Listerlaan
2. Schouwweg / Rustenvreugdlaan
3. Beukenhorst
4. Schouwweg / Oud Wassenaarscheweg
5. Beukenlaan / Oud Wassenaarscheweg
6. Wittenburgerweg / Stoeplaan
7. Langenhorst
8. Rust en Vreugdlaan
9. De Wittenburg
10. Duindigt
11. Rijswijk / Overvoorde
12. Te Werve and de Vredenburgweg
13. Ockenburgh
14. Willem de Zwijgerlaan
15. Zorgvliet
16. Haagsche Bosch
17-18. Kerkhoflaan (Scheveningseweg)
19. Rijswijksche Bos
20. Radio Sounding Station
 - building Bataafsche Import Maatschappij
  (RoyalDutch Shell)
21. Radio Sounding Station
 - block of flats at Jozef Israelplein
22. Waalsdorperweg
24. Staff housed at Hotel Schmitt
25. Haagsche Bosch firing sites (Jan. and late
  March 1945)
26. Bloemendaal
27. Technical Troops & Field Store Scheveningen
  (Nov. 1944)
28. V-2 Unloading Station
 - at Hartenhoekweg / Zwolschestraat

○ V-2 related sites
-------- Railways

in the operational areas. They wanted to compare the logistics of actual operations with the training approach given at the school. It was found that situations for supply in cities such as The Hague were far different that those of the batteries in rural locations. Scheveningen soon became the position of the technical troop's field store. The V-2 unloading station was located at the intersection of Hartenhoekweg and Zwolschestraat, while the assembly and transfer point for the V-2 technical and firing troops was located at Villa Bella Vista on the Scheveningen Promenade and also at the nearby underground CAP garage. Some of the personnel were housed in the opulent hotels nearby. The Dutch resistance reported that at least 14 Meillerwagen trailers had been seen outside of the garage. Staff headquarters were located further south at van Stolkweg, near the city park of Scheveningse Bosjes, where there was a launching site close to a small lake. V-2s were also launched from the streets Willem de Zwijgerlaan, Zorgvliet, near the intersection of Jacob Catslaan and Johan de Wittlaan, Kerkhoflaan, Waalsdorperweg, and on the road to Waalsdorp across the railway line to Scheveningen. All traffic would be cleared from the selected roadway just before launch. The rocket units moved to a new site every few days. A tree-lined street running through The Hague through Wassenaar was an ideal launching site, as long as it was firm and level, accessible to all launch vehicles, and provided natural camouflage for the rockets. The huge forested Haagsche Bosch park inside The Hague was a favorite launching ground.

Besides the firing sites at Bloemendaal and Ockenburgh, several miles to the south was a Battery 444 firing site at Hoek van Holland (Hook of Holland) on a new concrete road located a few hundred yards east of the railway station, opposite the car ferry that sailed between Hook of Holland and Rozenburg. The site was rumored to be controlled from a concrete bunker that had been built as part of the coastal defenses protecting the entrance to the New Waterway near the old fruit wharf. Perfect little squares were scorched into the new concrete road where the firing tables had been standing southeast of the station.

In November of 1944, a new unit was sent to The Hague: Artillerie Ersatz Abteilung 271 (Artillery Replacement Detachment 271). Detachment 271 was originally formed to act as a buffer for the operational V-2 units. Throughout the greater part of 1944, they were responsible for training the new recruits in V-2 field operations, while at the same time, their members were available to supplement the existing operational units. On November 23, 1944, Detachment 271 was incorporated into Battery 444, its additional members helping to finally bring the unit to full strength. Battery 444 now comprised three operational firing platoons, a technical troop section, a fueling section, and a target acquisition and radio section.

By mid-December, V-2 launch sites had been prepared at several new locations. These included a site near Kasteel (castle) Oud Poelgeest at Rijswijk, another at van Vredenburgweg Huis te Werve, and one near the former Rijswijk railway station. On December 16, rockets were reported to be launched between Egmond and Bergen aan Zee in the dune area. A few days later, the Dutch resistance reported that close to a hundred rockets were seen lying underneath the trees of Haagsche Bosch park.

In Britain, officials were looking for alternate methods that could possibly reduce the number of long-range missiles being fired. Intelligence officials drew up a list of 18 possible factories for manufacturing liquid oxygen. The list included ten locations in Germany and eight in Holland. Some suppliers in Holland were the companies NV Centrale Ammoniak Fabriek at Weesperkarspel, which delivered liquid oxygen exclusively for the German troops in The Hague; a company in Amsterdam known as NV Maatschappij tot explotatie der cg Remmenhullersche Koolzuur en Zuurstoffabriek Kerkstraat 271, which also delivered liquid oxygen; and two companies in Delft: Gist en Spiritus fabriek and De Destilleerderij en Roomgist fabriek, which delivered alcohol. The Air Ministry began an analysis to discover whether it was possible to bomb these factories. However, the eight plants in Holland, considered to be the most important, had been built in residential areas. Attacking these factories would require pinpoint accuracy; even the slightest bombing error might result in hundreds of civilian deaths. Because of an agreement with the Dutch government in London regarding such targets in residential districts, only one liquid oxygen factory in Holland—Zuurstof fabriek De Alblas—was bombed. On January 22, in the mid-morning, the plant was attacked by seven Spitfire fighter bombers, then again around 1:00 PM, destroying it. Only two of the German facilities were attacked, but because most of the vital plants were far underground, the raids in Germany had no effect upon the supply of liquid oxygen transported to the launching crews.

At around the same time, Allied armed reconnaissance flights over The Hague were stepped up, and hundreds of sorties were flown over the Dutch coast. The air-raid sirens sounded many times each day near the launch sites. From time to time, contacts in the Dutch underground tipped off the roving fighter bombers by providing information about recent German rocket activity. Between October and November, the Second Tactical Air Force flew more than 10,000 sorties against railways and road transportation targets between The Hague and Leiden and near Hook of Holland, destroying 40 barges, 40 locomotives, 200 railway cars, and over 200 motorized vehicles.

The very first Australian fighter wing to operate in Europe was a Spitfire wing, which included No. 451 Squadron and No. 453 Squadron of the Royal Australian Air Force. These fighter wings were devoted entirely to countering the V-2 rocket. Operating from bases in Britain and on the continent, the wing flew 1,328 sorties over Holland, bombing and strafing reported launching areas, workshops, and motorized transport and cutting railway lines leading to the firing sites. Pilots would sometimes receive heavy antiaircraft fire in random areas of The Hague and return later to investigate the possibility of V-2 activity; oftentimes flying multiple missions in a single day, attacking anything spotted moving on the ground. The air strikes were the only way of stopping, or at least slowing down, the rocket attacks. The mobility of the rocket crews meant that frequently only suspected areas of activity were bombed, as the rocket crews, after firing the missiles, had long since vacated the area. However, during this period, the primary focus of the air campaign was to destroy the overall German military effort in Holland, not the rocket sites specifically. There is no doubt that the Allied effort must have hindered the flow of supplies into Leiden and The

Hague in some manner. Especially when at the end of November, Allied fighters caught two trains in the open just from the Mittelwerk factory. Each train carried 20 missiles, and after the attack was concluded, all 40 missiles had to be scrapped. In spite of the many Allied attempts to stop the launchings, only rocket malfunctions and the occasional wayward shot kept the missiles from hitting their targets. An average of four or five V-2s struck London every day during the early part of December, with as many hitting the neighboring county of Essex just to the east.[10]

Members of the Dutch resistance learned very quickly that if a Meillerwagen trailer arrived in the area, along with the rest of the battery's vehicles, it was the sure sign of an impending launch. The rocket troops knew the underground was watching, because the fighter bombers came soon after each round was fired to attack the German positions. All along the Rijksstraatweg, the road north from The Hague through Wassenaar, foxholes had been dug for quick shelter in case of air attack. At one place along the road, a large sign informed pedestrians "Attention! Strafing attacks; foxholes on left." Dutch civilians were even subject to attack if they were in the wrong place at the wrong time. Soon the rocket crews realized that the launch vehicles were giving away their intentions. In December, a camouflage platoon was added to each firing platoon. Field training now emphasized camouflage for concealing the tankers and trailers from the air. Vehicles that remained in the launching areas were dug in and covered with concealment netting; the rest were dispersed and hidden in the woods.[11]

High-flying Allied bomber crews were also viewing this daily onslaught of V-2s rocketing skyward toward London and Antwerp. On November 24, while flying at 22,000 feet, bomber crews of the 544th Bomb Squadron got a close look at V-2s being fired at London. The B-17 and B-24 crews could easily see the fast-moving missiles and their contrails against the blackness of the upper atmosphere. On December 6, 1944, several Spitfires of No. 602 Squadron RAF were crossing out to the north from The Hague after attacking V-2 positions at Wassenaar. Just past 11 o'clock in morning, the pilots saw twin contrails rising from the Den Helder area. So there is some evidence, at least visual evidence, from the RAF pilots patrolling over the rocket sites that in December another launching site existed far north of The Hague, but at this point no official German record has been discovered that positively confirms this.[12]

On November 3, 1944, the second battery of 836 fired a final V-2 against Antwerp before they were ordered back to Peenemünde to test the possibility of launching V-2s from special railway equipment. While other batteries continued the rocket offensive, they would remain in the east until January 23, 1945.

On the dreary evening of November 11, 1944, a V-2 slammed into Shooters Hill in Greenwich, creating a ghastly scene. Striking a road that was filled with commuters, it left a massive hole in the roadway and consumed everything in its path. The passengers on a city bus were set ablaze when the flash ripped off the top deck and fire engulfed the remainder of the vehicle. Across the street, a local pub collapsed, crushing and then incinerating all of the customers inside. Unlike the V-1, the V-2 blast effect had a tendency to set on fire everything in the immediate area. Resembling some

One of the reasons it was very difficult to locate and attack the V-2 launching positions was the extensive use of camouflage by the rocket batteries. Special camouflage teams were responsible for making sure the sites were well hidden from Allied aircraft. This Hanomag SS100 tractor is draped with standard camouflage netting, which was a requirement for every vehicle of the battery. (*Deutsches Museum Munich*)

scene from Dante's *Inferno*, ruptured gas mains were observed belching huge bluish flames in the air as rescue squads dug away in the rubble.[13]

On November 25, 1944, while a V-2 was being tanked up in Wassenaar, people all over London were enjoying a nice shopping day. In West End, American soldiers could be spotted among the bustling streets. In the East London district of Deptford, at the New Cross Shopping Center, it was as crowded as ever on that Saturday afternoon. Shoppers were packed into Woolworth's, busy choosing from the store shelves. At 12:10 PM, following a blinding flash, the Woolworth's building was shaken apart by a massive explosion. An instant later, the entire building collapsed into the basement. Everyone inside was thrown down into the cellar along with tons of beams and plaster; many were buried under the huge pile of debris as bits of material continued to flutter from the sky. One hundred and sixty-eight people died. Some were killed outright by the rocket's impact; others were crushed or suffocated when the building caved in. Nearby there was again a city bus, its windows and tires shattered, with rows of people all covered in dust and sitting very still, all dead.

People in London were allowed to walk near the destruction, passing first houses with windows broken and roof tiles missing, then ones with outside walls missing where furniture could be seen inside. Further on there were just piles of bricks, the smashed rubble, and a huge hole in the road revealing broken pipes and cables.

At the beginning of December, a rocket came down in the River Thames, not far from London's Savoy Hotel. It sent a huge geyser of muddy water high into the air and blew out windows all up and down the riverfront. Later that day, Londoners gathered on Waterloo Bridge and the river embankment to stare at the spot where the rocket came down. Still, no official word had been given as to what was causing the bangs, but Londoners suspected they knew anyway.

"There are always two explosions, it seems," Mrs. Gwladys Cox of West Hampstead wrote in her diary. "The things travel through the stratosphere in two parts. One, the container, explodes on reaching our atmosphere. This releases the bomb itself, which drops, digging a deep crater and pulverizing everything in its way." Mrs. Cox may not have understood the mechanics of Hitler's latest weapon—though, her analysis may not have been too far from reality, considering the airburst problems—but it is evident from her diary that people were talking.[14]

Besides causing actual physical damage, the rockets had a great psychological impact on Londoners. The abruptness of the V-2s arrival was its most terrifying feature. There was no time to do anything. This new affliction was much more supernatural than the flying bombs had been earlier that year; no one expected the "lightning bolt out of nowhere." Even though the V-2 was generally more feared, it disrupted everyday life less than the V-1 simply because there was no way to prepare for it. V-2s came unannounced, so most people just went to bed at night and prayed that they would be there the next morning. During the night they might be suddenly woken by a boom, very close; looking around, they'd find themselves covered with glass and ceiling plaster. After hearing the bang, people could often see where the rocket had fallen by a column of smoke that would rise and hang in the air as little bits were fluttering down. Whenever there would be flash of light, people were prepared for the boom of the explosion to follow, another "bloody rocket." The V-2 impacted at a speed that would bury the missile 30 feet into the earth before the warhead could explode. The blast wave, combined with the speed of impact, would devastate everything within a quarter-mile radius. The lateral blast damage was not as great as the V-1, but later it was discovered that many buildings that were thought to be intact had actually sustained substantial structural or foundation damage caused by the shockwave of impact. Londoners reported that without question, the V-2s were worse than the V-1 attacks. The thunderous bang of the explosion, followed by the ominous roar of the rocket's descent, made it seem bigger and more frightening.[15]

On the afternoon of November 16, 1944, the Dutch residents of the small community of Hellendoorn in middle Holland noticed unusual German troop movements among the dense trees of the Eelerberg Forest. The ground was still wet from recent rains as dozens of German soldiers appeared unexpectedly in the middle of the night. SS Werfer Battery 500 had arrived. Under the command of SS Captain Johannes Miesel, the unit comprised about 400 soldiers of the Waffen SS, specially trained for V-2 operations. The convoy of approximately 100 vehicles split into two sections. One detachment moved into Hellendoorn, and the other moved north to the forest at Archem. The

Damage to London suburb caused by a V-2 strike. (*NARA*)

troops of the measurement battery calculated the launch site coordinates according to the triangulation procedure. Seventeen Dutch families, all within one kilometer of the launching areas, were evacuated, and the SS soldiers then immediately occupied their homes. A command center was set up in Hellendoorn at the local parsonage.[16]

Hurriedly, the troops busied themselves in the distance, their breath condensing in the cool autumn air. Some type of giant projectile, the center of their attention, stood menacingly at the edge of the forest. After the rocket had been transferred to the firing troops from the technical troops at Archem, it had been erected on the firing table. This was followed by a parade of various type trucks and trailers driving up to the rocket, and then only silence from the forest. Thereafter, a strange armored halftrack vehicle—the Feuerleitpanzer firing control vehicle—noisily clattered up to the launching site. Around 3:00 PM, a series quick, heavy beats was heard, followed by a tremendous roar. It was unlike anything heard before in Hellendoorn. The rocket rose slowly from its perch, spewing fire and smoke. It came up several meters, then twisted over and crashed into the wet earth not far from the launching point. The resulting explosion shattered up to 150 windows of a nearby sanatorium hospital, sending splintered glass fragments raining everywhere. Dozens of medical personnel inside the building were sent scrambling for cover.

For the Dutch counties of Twente and Salland, this was their introduction to the V-2. The Hellendoorn area was about to become linked in history with Hitler's vengeance weapon. The sight of the huge torpedo-shaped projectiles rising from the forest, with an even longer mass of fire coming from the tail, followed by the long twisting contrails of white smoke high in the heavens, soon became a common occurrence for the citizens of Hellendoorn. Beginning on November 17, V-2 rockets were fired several times daily, the target being the Allied-controlled port at Antwerp.

The SS Werfer Battery 500 was broken down into the usual elements of firing, supply, and technical units. There were other common sections such as a headquarters office, a first-aid station, and a field kitchen. The battery possessed about 100 vehicles of varying types. The firing battery was divided in three Schiesszüge (launching platoons). These again were divided into specialized groups; some men handled rocket propulsion while others maintained the electrical installations, including the Feuerleitpanzer. The fuel trucks and trailers were not part of the firing battery but belonged separately to the supply battery.

Because the unit was the only SS battery of the division, they received special attention of the division commander, SS General Hans Kammler. Repeatedly, he wished that the SS 500 would become the battalion with the highest average number of daily launches. But it was the Army's Training and Experimental Battery 444, firing against London from The Hague that achieved the highest rate. Kammler also made sure that the SS 500 received the best in equipment and training. Like the Wehrmacht V-2 battalions, the personnel received their training in Köslin. However, the SS troops, because they were perceived as a special battalion, received the extra instruction at the underground factory at Nordhausen. They were allowed access to missiles under construction, and Peenemünde engineers gave further instruction in greater detail about the rocket's mechanisms. None of the other V-2 crews had been given this extraordinary education at the rocket factory. This was Kammler's decision, and it demonstrated that he wanted to give his SS unit an advantage over the other battalions, all of which belonged to the regular Army.

Heinz Wellmann, in the third firing platoon of the SS 500, knew nothing of Kammler's expectations. He was a young man who suddenly found himself on the western front, in the middle of a Dutch forest, with the perilous business of launching V-2 missiles.

"We were never told exactly where we were. We guessed we were in Holland, somewhere near Belgium. We were so busy we had no time to look around. The house we slept in was not far from the launching sites, and was much better than the tents we had at Heidekraut! Our third platoon normally just walked to our firing position and back. Once in awhile you could go to the village, but mostly we had no time. It was not prohibited to contact the Dutch people, we just never met them. In the daytime the Dutch could go along the main roads unrestricted, but they could not enter the guarded areas where we were. They had no motivation to come in contact with us anyway."

At Hellendoorn there were at least six firing sites, maybe a few more. They were located in the Eelerberg forest about one mile northwest of the town. The forest, which was entirely coniferous, was crossed by a number of roads and small paths. All firing sites were close together on or beside these pathways. Some of the actual firing plat-

The Feuerleitpanzer (firing control vehicle), Zugkraftwagen (heavy traction vehicle) Sd.Kfz.7/3, was built on the popular eight-ton German halftrack chassis. It operated as control post for rocket launching units. Mounted on the rear of the Feuerleitpanzer was an armored superstructure to house the guidance and radio equipment and also to protect the crew of three men from the blast and/or explosion. During launching, the Feuerleitpanzer was positioned about 100–150 meters away from the rocket, usually down in a protective pit. Two operators sat at the firing control console or steering table (right), while the launch control officer stood behind them. The control operator could see the launch area by opening a hatch in the roof, but at the moment of firing, the hatch was closed. In front of them were protective vision slots through which they could view the rocket on its launch table. (*Deutsches Museum Munich*)

forms were built on pine logs cut to equal size, trimmed, and bound together with wire; however, a few firing points were only leveled sand. The security perimeter for the firing position included all of the surrounding forest. The entry routes into the forest were guarded by sentry posts.

"When we arrived, our launching sites had already been cleared and prepared for us," remembered Wellmann. "We only needed to drive in with our firing table and plug in our connections. We situated and adjusted the table exactly; we could adjust each leg separately. At some of the positions we used a three-centimeter-thick steel plate under the table to stabilize it, otherwise the engine exhaust would have blown away the soil under the table. Even so, part of the soil was still blown away each time, but not so much."

The battery officers were quartered at House Eelerberg, a large estate north of the firing sites, while the firing crews were housed in farm buildings near the sites. The second firing platoon was billeted in a house directly across from the launch sites with their vehicles parked behind the house when not in operation. During the launching periods, men arrived day and night, and there was a constant coming and going of people carrying boards and map cases and bringing in reports.

Sixty years later, Wellmann remembers: "Sometimes we fired from the same position as the platoon before us, sometimes not. But we always brought our table and Feuerleitpanzer with us. Each platoon only had to pull out their connecting cables, they looked like big electrical sockets, and then the platoon could move away for the next. To move the site's fixed cables was a big job; they had to be rolled up, and that was done by the survey troops. Those cables were quite long; they couldn't be transported inside the launch control vehicle. When we had to change firing positions, we had to help to disassemble all of our equipment, then the halftrack picked up the table, and then we went away to our next position. The platoons would fire night and day in shifts. When one platoon went to the launching site, the other came back. Our firing crew never mingled with the other platoons. It was rare for us to meet a soldier from another firing crew, you may have spoken with one from time to time, but we didn't get to know most people of the other platoons. When the battery had group meetings, we were all together, except for the drivers, but you did not know who everyone was or what job they performed."

Rockets bound for Hellendoorn were transported from Germany by rail and delivered to Nijverdal. At Nijverdal, the unloading area was located just west of the station, on the hill leading out of town. Dutch residents in Archem reported that rockets were transferred to the field store near the woods between Marienberg and Ommen. Two long sidings existed to the south of the main line running from Marienberg to Ommen; these sidings served as a main transfer point for rockets.

"The Meillerwagen would then bring the rocket. It was covered in a canvas tarp; a nice thick tarp, so the rocket would not get wet. They were zipped up and divided into two parts; one part covered only the tail. The rocket was raised with the tarps in place; we only loosened them a bit. The rocket was resting on the big arm, in the middle were bands, and up at the nose was a kind of clamp. The Meiller trailer had a Volkswagen motor to drive the hydraulics, and when the V-2 was tilted upright we loosened the clamp somewhat. Once the rocket was standing on the table, the main strap in the middle was loosened. This was a steel band, on the inside it was padded with felt so it would not push into the rocket. Next the Meiller was pulled back a little, and then the tarps came down. Before we began fueling, some preliminary tests were run. At the bay compartment there were large, four-sided plugs connected into the electro-housing. The rocket was switched through

The Steyr 2000, 4x4 Heavy, Stromerzeugewagen (power supply vehicle), was the electrical heart of the V-2 launching site. This modified vehicle supplied power to the rocket on the firing table during the pre-launch tests and preparations. The onboard Zundapp engine-driven generator set delivered 220/380 V 4KVA, while the transformer delivered from 380 volts AC to 27 volts DC. It supplied the necessary types of voltage to the rocket and its various test circuits and also allowed for the conservation of the onboard batteries. (*Deutsches Museum Munich*)

electrically from the Feuerleitpanzer, and then the tanks were purged while empty. We just observed and waited until the checks were finished. After that, time was running. Fueling had to be done quickly."

The unloading point for fuel supply was the station at Heino, although the station at Ommen was sometimes used. After filling their tankers, the fueling vehicles usually went to the presumed launching site but occasionally returned to the parking place in the forest for a short time before going to the launching site. A launch would occur a few hours following the departure of fueling vehicles to the launching sites.

"Our launching crew consisted of about 20 to 25 fellows. Each of us was trained in certain tasks which we performed. Some worked near the bottom of the rocket, while I worked at the mid-section with a few others, and at the top were those dealing with the control and electrical systems. We worked as a team, each on our individual function, so nobody interfered with the other. The arm of the Meiller trailer had ladder rungs welded to it so we could climb to our work platforms. There were also built-in pipe connections to aid in fueling the missile. The hose from the fuel tanker was connected to the pipes on the Meiller, and up at the mid-section there was a hose which I connected to the rocket for alcohol fueling. First I had to remove the hatch covers, which was difficult because the screws were very small and easy to lose. Once the hatch was opened, I had to push aside the fiberglass insulation and free the tank connectors. Then I would connect the alcohol hose. The whole fueling process normally took about half an hour. The man working up top with the electronics could only close the bay compartment doors with permission from the launch control officer in the Feuerleitpanzer. We normally came down only minutes before the launch. On many occasions I walked away from the rocket only two or three minutes before firing. The man who started and inserted the igniter was the last to run away. Several times we experienced some type of unknown problem after the missile was already fueled and ready to fire. When the error could not be found quickly, we had to bring out these large hot air blowers. They were positioned under the rocket to blow hot air inside the rocket to prevent the liquid oxygen from freezing up the internal ducts. These blowers also were used at times when launch preparations took too long. If we couldn't find the caused of the problem, the rockets had to be defueled and sent back. This was a real headache, but it only happened to us a few times."

Cor Lulof was a nine-year-old boy at the time of the V-weapon activity near Hellendoorn. During these months, a normal child's time would have been occupied with school, but in Holland during the occupation, things were different. The Wehrmacht used Cor's school as a barracks, so near Hellendoorn, parents taught their children at home. Leather was not available, so they wore wooden shoes. Food was

The Pfaff manufactured 3,600-pound Abschussplattform (launching table) was sturdy enough to support the weight of the 12-ton, fully-fueled V-2, yet was highly maneuverable when mated with its custom towing dolly. The legs of the table featured built-in jacks to adjust the height of the table's rotating ring. Once the legs of the table were screwed down, the towing dolly was rolled away and the rocket was erected above the table. The table was then raised until the fins of the rocket were resting on the four support pads of the rotating table ring. The center of the table incorporated a blast deflector with a hardened steel tip to divert the exhaust gases away from the bottom of the rocket at ignition. The support brace for the power cable mast was welded to the rotating ring, as were the supports for the oxygen topping off supply and the five-way coupling, which were used during preflight tests. (Author)

rationed; in fact, everything was rationed, and the Wehrmacht confiscated all leather, nonferrous metals, horses, bikes, grain, cattle, radio sets, etc. Electricity was cut off, so at night there was only candlelight. Windows had to be blinded against air attack, and anyone caught on the street after 8:00 PM without a permit was shot.

Cor's father worked for a firm related to the German Army and managed to avoid the Nazis who were rounding up slave laborers. By using fake permits to confirm his important work for the Wehrmacht and to keep his bicycle, Cor's father was not removed from his family. Cor also owned a bicycle, too small to be confiscated, so he joined his father every weekend on a quest for food. Food was traded for wool, stolen from the German firm he worked for; in those days, wool was a precious commodity.

Cor's uncle had a son named Henk, who was a bit older than Cor. While their fathers talked about the war, drank Ersatz coffee, and smoked homegrown tobacco, the two boys would roam through the forest in the direction of the sanatorium hospital, which was a prewar fresh-air center for tuberculosis patients. Late in 1944, between the Dutch holiday of Sinterklaas and Christmas, Henk told Cor a confusing story about big guns, aircraft, and lots of smoke and fire, seen on occasion half a kilometer to the north. Naturally, the boys started across the wooded terrain in that direction. They did not follow the road west to the sanatorium because the intersection was guarded by German troops and there was occasional German traffic on that road. Walking through the woods, they heard a horn or claxon signal some distance ahead of them, followed seconds later by a

V-2 on firing table with final adjustments being made. The arm of the Meillerwagen contained many accessories for accommodating electrical tests, fueling, and servicing of the rocket. V-2s manufactured in the last months of the war were painted olive green over all surfaces. (*NARA*)

tremendous roar, then a big cloud of smoke could be seen rising above the trees. The roar increased to a thunder; not only did they hear it, they could feel it pounding in their chests. From the cloud, a big, dark, gray-green burning thing, the size of a church tower, rose into the air trailing a huge flame on top of a pillar of gray smoke. They scrambled away as quickly as possible, shocked and stunned by the experience. Running all the way back to the farm, they only stopped for a moment to look up. In the sky they could see a strange, kinky, broken contrail. From his home in Almelo, Cor had seen this many times before without knowing its origin. It would become known as "frozen lightning" to those who witnessed it so many times.[17]

Early in December, shortly after liftoff, a rocket fired from Hellendoorn malfunctioned and plummeted to the earth about five kilometers to the southwest, near the small town of Luttenberg. Dutch civilians witnessed the crash and hurried to the impact site to get a look at the curious machine. In the days previously, many launches were seen from a distance, but no one had ever viewed the rockets up close.

"It was in the afternoon on December 4, 1944. I suddenly saw an incredible sight of this thing falling from the sky," recalled Mr. A. Kleine-Toereers. "I wasn't sure exactly what it was, probably a German rocket. My neighbor came over on his bicycle and we went together to see. After we had ridden some distance, we could finally see the big hole in the land where the thing had fallen. I placed my bicycle near the ditch and walked some ten meters across the field and there I looked into the deep hole. I could see a fierce flame burning. Many others gathered around the rim of the crater to view the object. Men, women and even children were standing very close, when suddenly there was an enormous bang and then we were all pushed to the ground. Perhaps, for about ten minutes, I laid there unconscious between the dead and dying. I was still dazed when I moved away. I could not hear anyone groaning, as they might, because I was totally deaf. I took my bicycle and was on my way for only a short distance when I met two acquaintances. They told me that my clothes were still burning at the top. I looked down to see my trouser legs were gone, I was wearing only the top portion of my pants."[18]

Nineteen people died at Luttenberg that afternoon when the warhead of the V-2, which remained unexploded after the crash, detonated with the crowd surrounding it. Windows on farm houses half a mile away were blown out. Everywhere there were bodies. A local nurse arrived on the scene about the same time as some German doctors. The doctors immediately started to help the victims. Some of the injured hurried away before the doctors arrived because they were afraid of the Germans.

In Antwerp the recipients of the SS 500's missiles had no idea where they were coming from. It seems the SS 500 firing crews were also in the dark. "Our superior officers never told us where our rounds were going. We could guess sometimes; if we observed the measuring-in of the rocket and then turned the V-2 on the dialing ring of the launch table, we noticed that one round was aimed in one direction and the next round in another direction. We knew rockets were being fired at London, but we assumed our rockets were heading for Antwerp, although it was never told directly. We only could orientate ourselves when the device was properly aimed. The measurement crew knew, but we were not told," said Heinz Wellmann.[19]

The V-2 launches occurred regularly at Hellendoorn until a break at the end of December. Total rockets fired up to that point was 126, with 15 rockets crashing during or shortly after liftoff. In late December 1944, the SS 500 pulled out of the Hellendoorn area. One of the reasons was because the neighboring V-1 launching ramps were too close. Flying bomb crews were protesting because the V-2 trajectories crossed right over the V-1 launching sites to the southwest, while the rocket crews had observed several flying bombs crashing near their launch sites. It was also suspected that the rocket failures might be noticeable to Allied aircraft, seen as large scars in the forest from above, and possibly revealing the firing locations. In addition, a Dutch resistance member was arrested by the German patrols, and in his possession they discovered a complete set of drawings describing the launch positions in Hellendoorn. Not knowing if more drawings like these existed or if they were already in Allied hands, it was deemed prudent to move the firing position of the battery further to the north.

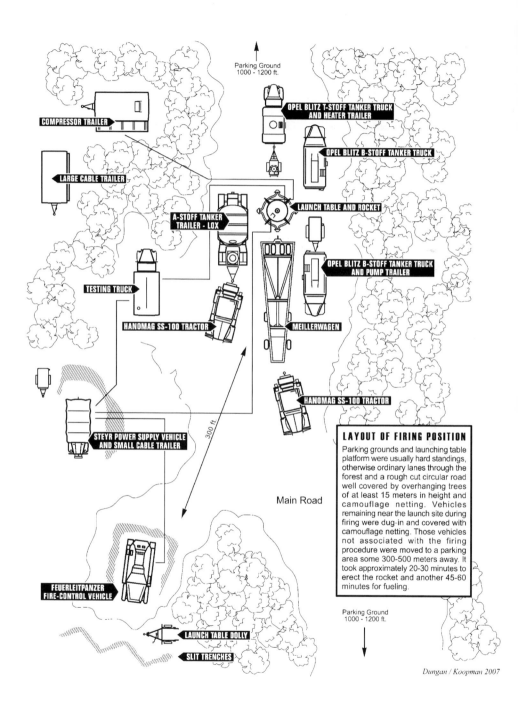

Parking Ground
1000 - 1200 ft.

COMPRESSOR TRAILER

OPEL BLITZ T-STOFF TANKER TRUCK
AND HEATER TRAILER

OPEL BLITZ B-STOFF TANKER TRUCK

LARGE CABLE TRAILER

A-STOFF TANKER
TRAILER - LOX

LAUNCH TABLE AND ROCKET

OPEL BLITZ B-STOFF TANKER TRUCK
AND PUMP TRAILER

TESTING TRUCK

HANOMAG SS-100 TRACTOR

MEILLERWAGEN

HANOMAG SS-100 TRACTOR

STEYR POWER SUPPLY VEHICLE
AND SMALL CABLE TRAILER

300 ft

Main Road

LAYOUT OF FIRING POSITION

Parking grounds and launching table
platform were usually hard standings,
otherwise ordinary lanes through the
forest and a rough cut circular road
well covered by overhanging trees
of at least 15 meters in height and
camouflage netting. Vehicles
remaining near the launch site during
firing were dug-in and covered with
camouflage netting. Those vehicles
not associated with the firing
procedure were moved to a parking
area some 300-500 meters away. It
took approximately 20-30 minutes to
erect the rocket and another 45-60
minutes for fueling.

FEUERLEITPANZER
FIRE-CONTROL VEHICLE

LAUNCH TABLE DOLLY

SLIT TRENCHES

Parking Ground
1000 - 1200 ft.

Dungan / Koopman 2007

In November of 1944, one the SS Werfer Battery 500's associated units—the Leitstrahlstellung (LS) guiding beam soldiers—took part in controlled shooting trials in Poland. The LS station for these trials was located some 1.5 miles west of Linsk and remained there throughout the shooting trials. The first eight to ten shots fired in Poland were directed to fall in the open spaces of Poland near Brenslau. Afterward, the trials were suspended for a few days while the impact areas were located. The remaining four to five test shots were guided toward the bay of Danzig with one overshooting and landing in Sweden.

In the dead of night on November 30, 1944, the first battery of 836 moved across the Rhine River from Hermeskeil to the Westerwald frontier. New launching sites were established near the German towns of Hillscheid and near Gehlert. The move was completed by the evening of December 1 with the first battery in positions near Hillscheid; they were followed by the third battery, which occupied the newly prepared sites outside of Gehlert on December 9. That same day, the first battery fired one rocket, while the third battery fired four. One of these rockets hit Antwerp-Kiel, killing 43 people and injuring 94, while another crashed into the Meir-Kathelynevest (Maritime Agency building).

Things did not go so well the next day. The first battery fired two rockets against Antwerp. The first rocket was a Kurzschüsse (short shot), which crashed into the road some 200 yards away from the launching site. A very large crater was torn in the roadway, and as a result, operations at the two other firing platforms were delayed. It was not possible to deliver the liquid oxygen because of the crater, which also blocked the path of the Meillerwagen to one of the pads. The second rocket did not come up to full thrust at liftoff. It tilted on the firing table and toppled onto the firing platform. The five-fold connector had not been detached from the rocket and caused the rocket to be pulled to one side. It was decided to drain the remaining fuel from the crumpled rocket by using a pickax to puncture the tanks. The draining fuels ignited and exploded. Three soldiers were killed and two others wounded. This was a great loss of material, and it also ruined the firing position. After the day was over, that particular platform was never used again. In total, 189 V-2s were eventually fired from Hillscheid with 23 failures of various causes.[20]

In December, heightened shortages of liquid oxygen were hampering the operations of Group South. This was mainly caused by the abandonment of the Wittringen oxygen plant during the Allied advance on December 7, 1944. The first and third batteries fired on average only one V-2 every three days. By December 15, 1944, the third battery of 836 was operating in the forests near Gehlert firing against Antwerp. There were at least five to six launching points at Gehlert; however, one site was completely destroyed on December 18 when a rocket fell back onto the firing table and exploded. From Gehlert the third battery would eventually launch a total of 199 V-2s over a period of the next few months.

In late 1944, not only was the V-2 in operation from the Westerwald but it was also being partially manufactured there also. Both V-1 and V-2 parts were produced along the small Dill River Valley, where there were many heavy iron industries, as part of the Nachbauring-2 production ring. The towns of Dillenberg, Niederscheld, Merkenbach, Eschenburg, Sinn, and Ehringshausen all had factories to build parts and pieces for the V-weapons. Burger Eisenwerke at Herborn painted some portions the rockets in their camouflaged color. (In this period only the ragged camouflage pattern was applied to the rocket parts.) At Wetzlar, another factory produced automatic pilots and controls for aircraft and also controls for the V-1 and the V-2.[21]

Fires rage in the streets of London following a V-weapon attack in 1944. The blast effect of a V-2 could devastate a whole city block, killing or injuring many and leaving hundreds homeless. (*NARA*)

By the end of October, Allied leaders realized that the Germans were directing V-2 fire at the vital supply port of Antwerp. General Eisenhower asked the British Air Ministry to transfer present Crossbow information to Supreme Allied Headquarters in order to create a continental Crossbow organization responsible for gathering intelligence to counter the German campaign against Antwerp. On November 9, 1944, following a hard-fought nine-day battle, the German commander at Walcheren surrendered the peninsula to British and Canadian forces; more than 10,000 German troops were taken prisoner. The way to the port was now clear; Allied forces controlled all 54 miles of the port's seaward approaches. However, the rocket attacks against London went on from The Hague and its outskirts, while the port of Antwerp was under fire from sites in middle Holland and Germany.

Hitler knew the importance of Antwerp. He knew the war was swinging in favor of the Allied armies and demanded his generals to devise a plan to slow the Allied advance. The Crossbow contingent in Belgium was just getting started when on the morning of December 16, Hitler's offensive was launched. Three German Armies, totaling 25 divisions, struck six unsuspecting American divisions in the Ardennes and overran their lines. Gallant fighting by survivors secured the preservation of two critical road junctions, and swift reinforcement by both American and British troops halted the German advance. However, the Battle of the Bulge saw the appearance of a whole host of new German weaponry. Crossbow officials were now pondering new threats such as turbojet and rocket-powered fighters, turbojet bombers, and other unusual technologies. There were rumors floating around that soon Germany would

Arrangement of Division z.V. operational units at the end of December 1944

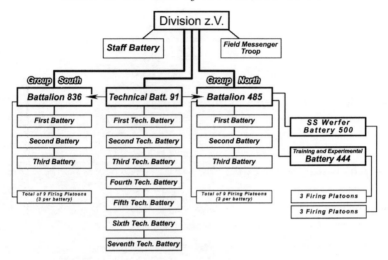

Command Structure of Division z.V.

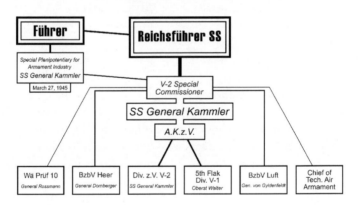

unleash an intercontinental ballistic missile capable of striking the United States. At the time it seemed possible, if not probable.[22]

On December 9, 1944, the leading figures related to the development and production of the rocket were summoned to Burgsteinfurt in the west. Near Coesfeld, in Nordrhein-Westfalen, the attendees gathered at the castle estate of Schloss Varlar in Rosendahl-Osterwick. The shadowy castle, along with the dark forests surrounding it, provided a truly gothic setting for the gathering that winter evening. However, the dinner was probably anything but gloomy. The posh dining hall was the scene of indulgence for the influential men responsible for Germany's latest miracle weapon. Albert Speer presided over a ceremony in which Wernher von Braun and Walter Dornberger were awarded the Ritterkreuz zum Kriegsverdienst-kreuz mit Schwertern (the Knights

Cross of the War Service Cross with Swords). Since von Braun was not in attendance, Dornberger accepted for him. Also receiving awards that evening were Georg Rickhey (director of the Mittelwerk), Heinz Kunze (V-weapons program director), and posthumously, the late Dr. Walter Thiel.[23]

Just one kilometer from the castle grounds was a launching site of the first battery of Battalion 485. A few kilometers further on were more launching sites, all in operation against Antwerp. The lights of the hall were turned down and the window curtains were opened. The attendees watched through the large windows as V-2s were launched from the forest in front of them, each one heading toward Antwerp. Following each launch, Speer would decorate one of the awardees. The bright light from the missile's exhaust would shimmer against the black of night, lighting up the hall for half a minute, as the rumble from the engine shook the hall. Twilight of the Gods—Wagner would have been proud.[24]

7

City of Sudden Death

The rumble of a lone V-1 could be heard above the fog like a dirge as the tiny procession made its way past the rubble and devastation. The high wheels of the horse-drawn hearse rolled slowly over the cobblestone street. The coachman on top guided the weary animals that were draped in long black blankets with thick black plumes attached to their bridles. Walking behind the hearse came a small group of mourners, all dressed in black. In the historic city of Antwerp, Belgium, it was a scene to be repeated many times during the long, cold winter to come.[1]

Antwerp lies 80 kilometers from the North Sea on the Schelde River. The river Schelde flows into the Dutch Schelde called Westerschelde. On the northern embankment of the Schelde lies Noord Brabant, then the South Beveland and Walcheren peninsulas. The port is one of the Europe's great harbors, but it is not a natural harbor. Its docks were dug out and fitted with locks to regulate the water and allow transportation of goods further inland, making it a great port city by design.

It was not until the last German forces were cleared from the heavily fortified stronghold of the Dutch Walcheren peninsula that the Allies were finally able to take advantage of the port's vast facilities. By November 1, after more than 27,000 casualties, Canadian forces had cleared South Beveland and Walcheren. It took another two weeks for the Royal Navy to clear the mines left by the retreating Germans in the estuary. Finally, on November 28, 1944, the port of Antwerp was opened. Unloading of supplies began immediately. It had taken almost three months to secure the harbor. However, it was worth the effort, for unlike the ports at Brest and Cherbourg, which had been completely destroyed by the retreating Germans, Antwerp remained surprisingly intact. Close to 9,000 Belgian civilians worked daily in the port unloading equipment and supplies with the Allied troops.

The V-2 rocket batteries firing on Antwerp were positioned at several places. In October, the first and third batteries of Battalion 485 and the newly activated SS Werfer Battery 500 were all firing from positions near Burgsteinfurt in Germany. The second and third batteries of Battalion 836 bombarded Antwerp from firing positions near the town of Merzig, and the first battery also targeted Antwerp from positions near Hermeskeil. The Training and Experimental Battery 444 also fired from Gaasterland toward Antwerp. The 444 was positioned in the wooded area near the northern Dutch town of Rijs. In November, all three of Battalion 485 batteries launched for a time from the Burgsteinfurt area against the Belgian port. SS Werfer Battery 500 had moved to new positions in Holland near Hellendoorn. All three batteries of Battalion 836 fired in unison from Hermeskeil until December, when they moved across the Rhine River to the area of the Westerwald frontier.

On October 7, 1944, a V-2 range-finding shot landed near Antwerp. It fell in the community of Brasschaat, about eight kilometers to the northeast without any casualties. A few days later, residents of Antwerp heard a tremendous explosion on the morning of Friday, October 13, when a V-2 rocket destroyed several buildings on the corner of Schildersstraat and Karel Rogierstraat. There were reports of many citizens being crushed under the tons of rubble. The infamous V-2 had just claimed its first victims in the Belgian port city. Later that same day, another rocket crashed into the city. The local residents came to the scene of the impacts for a closer inspection. Fears among the city's population were increased, but panic did not ensue. The local newspapers gave no real hint to the actual cause of the blasts, but they did urge the Antwerp residents to take certain precautions if they encountered the German flying bomb. (This is curious because in reality, no V-1 struck Antwerp before October 25 because of bad weather conditions and high winds.[2]) After four years of German occupation, it soon would become clear to the residents of Antwerp that the Nazi scourge on their city was not over—the German wonder weapons were about to target the historic port city.[3]

Greg Hayward was an 18-year-old airman with the Royal Air Force during the invasion of Europe. He served with the 146th Wing, equipped with fighter bombers supporting the British and Canadian forces during their advance from Normandy to the German frontier. Hayward arrived in Antwerp on October 2 and was based at Deurne, a prewar civilian airport just three and a half miles from the city's downtown. During the following days, Hayward and his comrades witnessed the increase of V-weapon attacks. On October 19, a V-2 rocket struck the Kroonstraat at Borgerhout, destroying 25 houses, killing 44 people and injuring about 100. The explosion on October 28 at Bontemantelstraat (one of the most densely populated parts of the city) was the first real massacre of the V-weapon campaign.

Hayward recalls a harrowing experience at Deurne: "The closest shave I experienced was on October 25 when a V-2 exploded on the airfield where, along with some 20 of my comrades, I was working on one of two aircraft dispersed for servicing. The incident left five airmen dead and a dozen or so injured. The rocket landed only 50 yards away but fortunately missed the concrete roadway and hit an area of soft ground—otherwise, I would probably be dead. I was standing on the wing of an aircraft at the time, waiting to climb into the cockpit. All I remember was a brilliant crimson flash, but no actual recollection of an explosion. Then, I was on the ground, the air around me black with smoke and dirt thrown up by the impact." Hayward recounted, "By the time I sorted myself out and realized that I was still in one piece, the smoke and dirt had dispersed, revealing a scene of devastation and a still-smoking crater where the rocket had landed. Rescue teams were immediately on the scene and ordered me back to our workshop to clean up. One of the dead airmen was a friend of mine, we had been at training school together, he was only 18!"[4]

Kenneth Hartman was a 23-year-old U.S. soldier stationed in Antwerp during the 175 days of V-weapon bombardment. His unit, Headquarters Q-189, Headquarters Company, 54th QM Base Depot, arrived in the city during October of 1944, finding neighborhoods seemingly deserted. Only a few children were seen in the streets. The job for his unit was to get the harbor open and operational. He remembers the V-weapons bearing down on the city as the symbol of the city, the bronze statue of mythical hero Brabo, held vigil in the old city center as the terror rained from the sky. "I would go to witness the result of the worst of the bombings. The V-2s did the most damage," said Hartman. "On October 14, I went to Schildersstraat the day after a V-2 had killed 32 and wounded 45." A rocket fired by the third batterie of Battalion 836 from Merzig came down at the corner of the Schildersstraat and the Karel Rogierstraat. Approximately 100 structures were damaged, including the Royal Museum of Fine Arts. Forty-three houses were totally destroyed.[5]

"A few days later, a V-2 impacted in the suburb of Borgerhout, at Kroonstraat, killing 44 and wounding 98. Three-story residences were literally sliced away from the next, leaving a common wall exposed to the weather," Hartman remembered. "I stopped going to the tragic sites after November 17. That day I came upon the ruins of the Boy's Orphanage at Durletstraat. The Belgian Red Cross had removed 36 dead children and another 125 wounded by a V-1 the day before." Also on November 17 another V-2 struck the St. Joanna Institute at Ferdinand Coosemansstraat. Thirty-two of the nuns died there under the massive pile of debris.[6]

In 1944, Charles Ostyn was an 18-year-old young man living in Hoboken, which is a suburb of Antwerp. He lived in a house with his parents and one sister. He was working in the city as a drafting apprentice at the time. Ostyn vividly remembers what it was like to live in Antwerp that winter: "On November 6, 1944, I got a phone call at the office in the city telling me to go home 'pronto' as a V-bomb had fallen in our street, but that was about all they could tell me. It turned out that some two hours before a V-1 had fallen into the park behind my house. The blast was not more than 30 meters from our house. It just so happened that a thick park wall shielded our home from the blast somewhat, the upper part and the roof was gone, but my parents were unhurt except for glass cuts. I spent the next few days clearing away the rubble, moving furniture and looking for missing things, but it was clear the place was uninhabitable. That winter was one of the coldest we had during the war and certainly the most miserable for us having lost our house early on."

To this day, Ostyn cannot forget the earth-shattering sound of the V-2s that rocked his city. "For the V-2 there were usually, but not always, two bangs separated by a split second, this I clearly remember. They were usually accompanied by a violent tremor if the impact occurred nearby. The approach of the rocket was only rarely observed." Amazingly, Ostyn actually witnessed a V-2 plunging to earth, "I saw this flash during the day, but only once, I just happened to look at the sky in the right direction. It was definitely not a contrail, but it was like a streak from a comet, as fast as a shooting star. It was a long, thin, white streak, more like a flash coming down to the earth. This was seen about one to two seconds before the impact."

"When a V-2 rocket hit in the city it was always followed by a huge black or brownish cloud of debris," said Ostyn. "If you dared to keep on watching, as I foolishly did many times, large pieces of metal and junk kept coming down all around you for several minutes. I always wondered if these twirling pieces of sheet metal were from the rocket itself. The only other time I saw a V-2 explode was on December 12; it was still dark outside and I was riding the tram to work. I didn't see a flash—just one hell of a bang and a yellow mass of flame lighting up the city. It hit about 500 meters away. I can vividly remember huge numbers of yellow streamers coming down with the shower of debris. These streamers were always fluttering down in bundles and must have been part of the rocket insulation."

Charles Ostyn was one of the few people in Antwerp who knew, on a daily basis, where the V-weapons had hit on any particular day. The office of his employer was on the fourth floor of a building on a narrow street in the old part of the city. Many times, after hearing the bangs, Ostyn would volunteer to climb to the roof and check on where the big brown cloud was billowing. He would stand in amazement looking at the junk and scrap pieces of metal floating down for several minutes. "It is one of the silly things you do when you are 18 years old I suppose."

After coming down from the roof, he would tell what he had seen and give the general direction of the impact. Everyone knew when a strike was very close because of the tremor, the falling of plaster, and the breaking of glass windows. An unnerving violent vibration could be felt in the building a split second before the two bangs.

Greg Hayward described the V-2 descent and impact this way: "One puzzling feature of the V-2 arrival was on a clear day, a descending vapor trail was clearly visible in the sky, and the first explosion occurred several thousand feet above the ground. This explosion appeared to be the rocket casing as hundreds of pieces of debris could be seen seemingly fluttering to the ground above the point of impact. I have never heard an official explanation of this phenomenon in the years since. All this, of course, was only seen after the warhead explosion alerted one to its arrival." In fact, what Hayward probably had witnessed was a midair breakup of the V-2.

On November 27, a terrible incident occurred at a major road junction near the Central Station. Teniers Plaats (Square) was the busiest intersection in town. Military policemen were always regulating the heavy traffic for an Allied convoy passing through the square. It was on the main north-south axis for the supply columns. From the docks, American troops were heading south to the U.S. supply bases near Mechelen, and British columns were heading north to the front lines in Holland. There were four tramlines crossing the square in both directions, plus there were many autos and pedestrians moving throughout the busy intersection.

"I often went there after lunch to watch the military activity," said Charles Ostyn. "There was a British MP, right there in the middle, regulating and directing both military and civilian traffic. On very busy days there were two MPs."

A V-2 came down at ten minutes past noon and exploded in the middle of all this activity. A British convoy was moving through the intersection and was caught in the blast. This particular rocket was believed to have exploded just above ground, possibly having struck the tramlines just where the traffic policemen stood. A city water main burst; water ran into the intersection. Soon, the whole square was filled with water.

"I heard and saw this explosion from a short distance away while riding in the back of an open truck and approached the scene about two hours later. There was water running everywhere and the whole place was cordoned off and guarded by U.S. soldiers. There was a massive crowd of onlookers and many people with bandages on their heads walking around. It must have hit something above ground first because no crater was ever found."

The result was total devastation. The water began to pool on the street. Floating on the water were corpses, various body parts, clothing, and large amounts of debris. Several of the vehicles in the convoy exploded or caught fire, and their occupants lay on the ground burning. The glass windows of the passing trams near the intersection were all shattered, causing injuries to those riding in them. One of the MPs was completely disintegrated, and the charred body of another was found some time later on the roof of a nearby hotel, about 60 meters away.

On that same bitterly cold November day, Simone De Ceunynck happened to be walking home from her place of employment. She worked as an assistant bookkeeper at a local insurance agency in Antwerp and was on her way home to Deurne for her lunch break. Simone had recently decided to alter her normal route because of all the V-weapon blasts occurring near the old path she walked before. Simone was in a hurry to get home, so she went ahead of her friends from the office. After walking about two blocks, she was approaching the city crossing at Teniers Square when suddenly she felt uneasy. She quickly darted in front of the Army convoy, trying to get across the street as fast as she could. She heard the loud voice of the nearby MP yelling at her just as she reached the other side of the street. It was at that moment that the rocket struck. All of a sudden the noise of the city stopped, there was a split second of silence, then a low rumble followed by fire and screaming. Simone found herself standing between many broken and bleeding bodies. The gloves she had been carrying were gone. Dazed, she looked around and saw them lying on a dead British soldier. Simone reached down to retrieve the gloves and was greeted with the awful sight of the soldier's fatal head wounds. She began to panic, screaming as the horror of the scene overtook her. Another British soldier calmed her down and escorted her to the nearest Red Cross station. Simone was bleeding but alive. Shrapnel had entered her leg and breast; however, she was one of the lucky ones. Although she did not know it until she returned to work days later, two of her coworkers had died in the attack.[7]

A well-known photograph of the devastation wrought by the V-2 impact at Teniers Square in Antwerp on November 27, 1944. The rocket came down at ten minutes past noon and exploded in the middle of a British convoy that was moving through the busy intersection. One hundred twenty-six were killed and another 300 were injured. (*NARA*)

First Lieutenant Verne W. Robinson just happened to be traveling near Teniers that day. Robinson was in a vehicle with Private Herbert L. Moyer and Private Marcel Snauwaert—all three members of the U.S. 604th Engineers. Driving from Namur, they were on a routine errand to a railroad station in Antwerp to pick up some supplies. As they approached the intersection at Teniers, they were slowed by the heavy traffic bustling through the crossing. In a flash, the mayhem of the rocket explosion left Lieutenant Robinson lying on the cobblestone street mortally wounded, as shrapnel pierced his temple. In the aftermath he was cared for by pedestrians, but to no avail. A tragic illustration of being at the wrong place at the wrong time, Lieutenant Robinson was the only soldier from the U.S. 604th Engineers to be killed during the war.[8]

Earlier, Greg Hayward was delighted to be off duty that day, and he was preparing to spend the afternoon in the town center. He was hoping to pay a visit to an American Red Cross canteen, where coffee and doughnuts were available. "At midday an explosion signaled a V-2 arrival, and it was apparent from the smoke and debris in the sky that it was somewhere near the center of the town," said Hayward. "After lunch I took a tram into town and as I walked from the terminal toward the Keyserlei, it was obvious that the site of the incident was very close. On reaching the area of Teniers Square, I saw a scene of utter devastation. My own lasting recollection is of street gullies running red as water from the broken main and fire hoses mixed with the blood of victims. Realizing that the area was closed and that there was nothing I could do, I returned to my base."

Teniers Square happened to be the geographical center of Antwerp. It is highly probable that it was the aiming point for many V-2 rounds. It had been one of the most traumatic days in Antwerp since the V-2 onslaught had begun. In all, there were 126 dead (26 were American and British soldiers) and another 309 injured. It was soon to get worse.[9]

Throughout November, V-1s and V-2s were punishing the city with more frequency. The death tolls were alarming high. The Americans were unloading approximately 10,000 tons of cargo each day, and by December had managed to almost double that amount. However, by mid-December the total amount of cargo off-loaded was in decline. After the beginning of the German offensive in the Ardennes, shipments began to stack up in the port. Material that normally would have been trucked to the front lines was held in reserve because of the fluid battlefield situation and for fear the material might fall into enemy hands.[10]

On the first day of the Ardennes offensive, December 16, 1944, Antwerp's worst V-weapon disaster occurred. The Rex Cinema on avenue De Keyserlei was packed full of people in the middle of the afternoon—nearly 1,200 seats were occupied—all watching the featured movie. At 3:20 PM, the audience suddenly glimpsed a split-second flash of light passing through the dark theater, followed by the balcony and ceiling crashing down during a deafening boom. A V-2 rocket had impacted directly on top of the cinema.

Charles Ostyn happened to be near the cinema that day and would later learn of a personal tragedy in his life caused by this particular rocket attack. "It all really sank in on us after the massacre at the Rex Cinema," said Ostyn. "I had walked past the theater about 20 minutes before the impact—to think, at that very moment a V-2 was being fueled, destined to soon kill all those people in one blinding instant."

The destruction was total. Afterward, many people were found still sitting in their seats, stone dead. For more than a week, the Allied authorities worked to clear the rubble, as 11 buildings were destroyed. Later, many of the bodies were laid out at the city zoo for identification. The death toll was 567 casualties, with another 291 injured. Two hundred ninety-six of the dead and 194 of the injured were U.S., British, and Canadian soldiers. This was the single highest death total from one rocket attack during the war in Europe.

"I heard the explosion while I was traveling home on the tram. The cinema was packed with more than 1,100 people and I remember the movie playing was *The Plainsman* with Gary Cooper and Jean Arthur; it was about Wild Bill Hickock—I was a real movie nut in my younger years. Later, I found out that my employer and his girlfriend were in the audience. Apparently, my boss, on a spur-of-the-moment decision, took his girlfriend out to see the film," said Ostyn.

Kenneth Hartman also learned that a U.S. soldier from his company had been killed in the Rex incident. "He had been with us for only a short time. A jeep load of us drove over to the site. The 350 Dispensary and 385 Engineer General Services Regiment were in charge of removing the bodies. I saw at least 600 dead bodies, still sitting in their seats, covered with fragments of plaster and concrete. A triage was in action to sort out nearly 300 wounded, most of whom were seated in the elevated rows above the many rows of corpses."

James Mathieson remembers the rocket struck the cinema just at the point in the movie where "Gary Cooper had captured an Indian who informed him that General

Destruction of the Rex Cinema on avenue De Keyserlei, December 16, 1944. Five hundred sixty-seven people died when a V-2 scored a direct hit on the crowded Antwerp movie house. Survivors said that the rocket came through the roof and exploded on the mezzanine. The rear row of seats in the balcony remained attached to the standing wall, and the wounded survivors were suspended above the wreckage and the dead. Rows upon rows of people were sitting in their seats, dead, as if still watching the screen. The rubble and debris was up to five meters high in some points, and it took the rescue teams six days to dig out all the dead. (Dr. Kenneth Hartman)

Custer and his troops had been wiped out." Mathieson was a member of an RAF intelligence unit, one of the first permanent RAF units in Belgium, which was stationed at German Admiral Erich Raeder's former headquarters in Antwerp. It just so happened that his CO decided he would allow a few men off to have a little break, and Mathieson decided to see a film.

Upon entering the cinema, Mathieson and a buddy decided to sit in the back row of the smallest portion of the balcony. When the roof fell in, Mathieson felt bricks and mortar falling from above. He put his left hand up to shield his head, which was quickly sliced open from the falling debris. Another brick landed on the opposite side of his head leaving a large gash. In a state of semi-consciousness, covered in dust and blood, Mathieson remembers being rescued from the debris.

"I was in a row where only three seats remained attached and I was lying over into space from the balcony. If I had gone down into the pit I would have had no chance. I consider to this day that I have a guardian angel looking after me because I think it was an absolute miracle that I escaped with so little injury."

Mathieson was moved to a British Army hospital in the Belgian town of Duffel. When he awoke a few days later, he discovered his wounds had been stitched up and his head and arms were wrapped in bandages. Amazingly, he was told the building housing his unit was hit by another V-2 the very next day and practically everybody was wiped out. Even though the V-2 explosion at the Rex almost killed the young Mathieson, the injuries he sustained may well have saved him from perishing with his unit.[11]

Survivors said that the rocket came through the roof and exploded on the mezzanine. The rear row of seats in the balcony remained attached to the standing wall and the wounded survivors were suspended above the wreckage and the dead. The rubble and debris were up to five meters high in some points, and it took the rescue teams six days to dig out all the dead. American and British teams had to join in with Army cranes and trucks. The hospitals were swamped, and the health services could not really cope with all the injured.

"The news that something really terrible had happened in the city filtered to the suburbs later that evening," recalled Charles Ostyn. "During the following week, it was finally confirmed that our boss and his fiancée were found dead under a thick layer of dust, both remarkably intact except for terrible head wounds. Thinking back, my closest call of being blown to eternity was one week after the Rex. We were at the funeral for my boss at Silsburg Cemetery at Deurne, and just before the coffin went down into the ground, at about 2:30 PM, a V-2 exploded at the other end of the cemetery, plowing into a row of houses, as if to underline the tragedy of it all. It was a very weird episode."

After this shock, all theaters and cinemas were shut down for a time and no more than 50 people were allowed to gather in any one place. People who could afford it left the city for safer parts, and Antwerp became a somber and semi-deserted city. The remaining citizens showed no real panic, but fear and tension existed in the city, especially after the Rex incident.

In 1945, *TIME* magazine referred to Antwerp as "The City of Sudden Death." Reporters had spoken to many of the U.S. soldiers working in the port area during the final week of the bombardment. The soldiers told of the terror rained on the city for the past four to five months. There was a total news blackout about the attacks in the papers, and this went on until April of 1945. What made matters worse was that this included any news about how the war effort was going. Any reports about locations of V-1 or V-2 hits would have given the Germans data that they could have used to improve their aiming. Therefore, the people of Antwerp never received any official information about what was happening. The V-weapon onslaught combined with the bad news from the Ardennes offensive in December made Antwerp residents realize that the war was far from over and that thousands more civilians and soldiers were going to die before Germany was defeated.

"The psychological effect on the citizens of Antwerp was great. It made us despondent and war weary, scared of what else Hitler had in store for us," said Ostyn. Between December 10 and December 16, about 761 civilians were killed by the V-weapons. The increased attacks in December could not have come at a worse time for the citizens of

Antwerp. The severe winter weather and the destruction of many houses added to the misery in the port city.

The only surviving piece of the V-2 was always the large combustion chambers, which were found all over the city and the suburbs, usually half buried in the ground. The chamber would sometimes careen after impact on the hard frozen clay and usually came to rest hundreds of meters from the point of impact, sometimes killing many people in its path. The combustion chamber weighed over 1,300 pounds. Ostyn remembers, "I was looking at one, half buried in the street at Groenplaats, a little square in front of the cathedral, it had traveled about 200 meters from the point of impact on Vrydagmarkt.

V-weapon activity was normally much less at night than by day. Still, there was the occasional ear-splitting bang late at night or in the early morning announcing the arrival of another 'Whispering Death' in the city. Only the loudest of thunderclaps can match this sound in its intensity or volume."[12]

By the end of 1944, greater Antwerp (city and port area, left bank and eight suburbs, population approximately 500,000) had recorded 590 direct hits, which had flattened 884 homes and caused around 1,200 others to be uninhabitable. Almost 6,000 buildings were badly damaged and more than 23,000 others were damaged in some manner. Casualty figures stood at 1,736 dead and another 4,500 injured. It seemed that no neighborhood escaped the destruction, as piles of debris could be seen everywhere.[13]

The coming New Year started no better than the previous one. Shortly after midnight on New Year's Eve, the city was struck by another flying bomb. Then, on January 2, the city registered no fewer than 20 V-1 strikes. The heavy snow made rescue work almost impossible. The municipal authorities had provided temporary housing for the bombed-out residents of the city, but they were almost overwhelmed by those in need. With the help of Allied forces, including doctors, nurses, and soldiers, they were able to cope. Rescue workers often had to deal with more than 50 corpses a day and the very real possibility that a wall or entire building, weakened by the blasts, might suddenly collapse on top of them.[14]

Off-loading of cargo continued, although at a slower rate, throughout December and into January. The problem for Allied transportation authorities was that the material wasn't going anywhere. Truck and railway transports were hindered to a great extent by the destruction to infrastructure caused by the falling V-bombs. Repairs to damaged rail lines, barge waterways, and bridges were being delayed because Belgian workers refused to enter the bombardment zone. Railway shipments were sitting idle while severed tracks were repaired. The V-weapons were also responsible for the loss of a considerable amount of Allied equipment; trucks, fuel depots, ships, and trains were all subject to the collateral damage from the intense bombardment. One ship was sunk by a direct V-2 hit, and 16 others were damaged at some point; the Kruisschans lock was damaged, several marshaling yards were hit, and the Hoboken petroleum installations were hit twice. Because of the V-bomb campaign on Antwerp, Supreme Allied Headquarters decided to divert all shipments of ammunition to the ports further south in France. This meant explosives would have to travel a greater distance to reach the front lines, further delaying the Allied advance and substantially reducing Antwerp's mission. A closer point of entry was needed, and in January of 1945, a new port was opened in Ghent. Activity at Ghent escalated rapidly. Ammunition could now be off-loaded without the fear of a catastrophic explosion caused by an inadvertent missile strike. In Antwerp, despite the bombardment, ships were still delivering supplies for the Allied war effort. Gradually, the flow of Belgian workers increased until thousands of dockworkers were unloading ships in the midst of V-weapon attacks. The civilians who worked at the docks received an extra bonus in their pay from the Allies. This bonus was called Bibbergeld—which literally meant "shivering money"—for the risks of working in the port while the V-bombs were falling.[15]

On November 10, 1944, American Brigadier General C. H. Armstrong arrived from Paris to take command of the Flying Bomb Command Antwerp X. This consist-ed of American, British, and Polish units of about 22,000 men. New plans had been drawn up for the three brigades of artillery to be stationed in a ring outside of the city to form a protective barrier against the flying bombs. Field Marshall Montgomery demanded that Antwerp X Command bring down half of all the V-1s launched at the city. This figure was reached in December (52 percent). In January 1945, they reached the 64 percent mark; and by February 1945, they managed to bring down 72 percent of all incoming V-1s. They had hundreds of guns spread out in three different gun belts around the town. Antwerp X would fire almost a million rounds of ammunition over the next few months and was responsible for shooting down more than 50 percent of the incoming V-1s.[16]

Many people tend to associate the V-weapon campaign as one directed only against England; however, Antwerp was the recipient of even more V-2s than London, result-ing in more than 30,000 killed or injured. For the whole of the V-bomb campaign, Antwerp received on average three V-2s per day in the city and its suburbs. The num-ber of V-1s was on average four per day in December and January, climbing to 12 daily in February of 1945. In late January and early February, the number of flying bombs had increased to the highest point and then tapered off in the month of March 1945. More than 1,600 V-2s fell on the port city during a six-month period. The V-weapon attacks on Antwerp came to an end as the German firing crews were forced to retreat because of the Allied advance.[17]

The V-weapon campaign against Antwerp is often overlooked by military histori-ans. The indiscriminate bombardment was certainly a terror for the civilian population of Antwerp, but it was also a monumental hindrance to Allied war planners. It is short sighted to say the V-weapons were ineffective simply because the port of Antwerp remained open throughout the campaign. Not even the Germans believed the rockets would completely destroy the port, but it was hoped, by amassing their fire on this strategic target, they could severely inhibit the Allies' progress toward Germany. In the weeks leading up to the Ardennes offensive, the V-weapons made it very difficult for supplies to reach the overstretched Allied lines. Hitler hoped to cut the American and

V-2 combustion chamber (left) photographed by American soldiers immediately after V-2 impact in Antwerp. These massive hunks of bent metal were usually the only recognizable piece of the rocket remaining after impact. Unidentified American soldier (right) poses with rocket debris in Antwerp. Items collected include large portion of turbo pump, combustion chamber, and the round tank for holding hydrogen peroxide, which along with calcium permanganate powered the turbo pump. (Dr. Kenneth Hartman)

British forces in half, with the capture of Antwerp being his ultimate goal. In the face of Allied air superiority, the V-bombs were Hitler's only available means to stem the flow of supplies prior to and during the offensive against Germany. Even though Hitler lost the Battle of the Bulge, the V-bombs continued to fall on Antwerp. Throughout the later portion of 1944 and well into 1945, the V-weapons severely curtailed the amount of supplies brought into Antwerp. The port never did reach its expected goals, and the Allies were forced to divert ammunition and manpower to Ghent.[18]

8

Final Retaliation

By December 23, 1944, the German counteroffensive in the Bulge was beginning to grind to a halt. They had outrun their supply lines; shortages of fuel and ammunition were becoming critical. The weather conditions over the Ardennes had improved, allowing the Allied air forces to attack. American heavy bombers flew devastating sorties on the German supply points in their rear, while P-47s started decimating German troops and vehicles on the roads. On Christmas Eve, German General Hasso von Manteuffel recommended to Hitler's Military Adjutant a halt to all offensive operations and a withdrawal back to the West Wall, but Hitler denied the request.

On January 1, in an attempt to keep the offensive going, the Luftwaffe launched Operation Bodenplatte, a major air operation against Allied airfields in the Low Countries. Hundreds of German aircraft attacked Allied airfields, destroying or severely damaging some 465 aircraft. However, the Luftwaffe lost 277 planes, 62 to fighters and 172 to the unexpectedly high number of Antwerp X flak guns, which unbeknownst to the Germans, had been set up to protect against incoming flying bombs. The loss of considerable amounts of aircraft on each side was more devastating to the Luftwaffe, as it could not replace the planes and pilots. The destruction of British aircraft of the Second Tactical Air Force was a blessing for the V-2 firing crews, however. The number of fighter bombers over The Hague was reduced for almost four weeks while the RAF struggled to replace the lost aircraft.[1]

On January 3, 1945, Montgomery's British 21st Army Group, along with the American First and Ninth Armies, moved against the Nazi armies from the north. They would meet up with Patton's advancing Third Army and cut off most of the attacking Germans. The Germans did their best to slow the Americans while letting their own forces retreat in good order. Much of the Wehrmacht escaped the battle area, although the fuel situation had become so calamitous that many German tanks had to be abandoned. On January 7, 1945, Hitler agreed to pull back most of the forces from the Ardennes, including the elite SS Panzer divisions. The Battle of the Bulge was now over; more than 80,000 American soldiers were killed, captured, or wounded; what remained for the Allies was mopping up the remaining German units. For their part, Germany's losses were 30,000 dead, 40,000 wounded, and 30,000 prisoners of war.

As Hitler's armed forces were hastily withdrawing back to the Fatherland, work continued at Peenemünde. In an effort to extend the range of the V-2, some 270 scientists were directed to develop an experimental winged version of the V-2, which was named the A-4b (the b stood for "bastard"). A wholly different winged rocket was originally slated to be called the A-9; however, the war situation did not allow for its continued development. The A-9 was based on the A-4/V-2 technology, and wind-tunnel experiments had already been conducted to determine the proper shape of the supersonic wings and tail fins. Portions of this design were incorporated into the A-4b blueprints.

The first test model of the "bastard" was launched from the beach area of Peenemünde on December 27, 1944. The rudder control failed about 100 feet above the launching point, and the rocket wobbled down the shoreline and crashed in the sand. On January 24, a second A-4b delivered better results. This was a vertical shot intended to test the sound barrier transition of the winged version. Liftoff and sound barrier transition succeeded perfectly, but on the downward trajectory one wing was ripped off the fuselage. It now clear that the A-4b would not have performed in glide mode due to insufficient power of the control system. This would conclude the A-4b testing.[2]

At the beginning of the New Year, in an attempt to increase his command stature, General Kammler reorganized all of the rocket units. Battalion 485 became the Artillery Regiment z.V.902. Battalion 836 became the Artillery Regiment z.V.901. The SS Werfer Battery 500 became the SS Werfer Abteilung 500. The Training and Experimental Battery 444 became Lehr und Versuchs Abteilung z.V. However, the Table of Organization, which upgraded the V-2 battalions to regiments, was not finalized until February 7, 1945; and even then, it was nothing more than a restructure on paper. The V-2 units were mobile and active—it was difficult to complete a reorganization without disturbing the launching activities. In the middle of January, the experimental grounds at Heidekraut were evacuated. The Russian advance, along with the dwindling supply of experimental rockets, prevented any further test shots. After a short stay in the frigid, snow-covered woods at Wolgast, the remnants of the Heidekraut organization were disbanded near the Weser at Rethen.[3]

On January 12, 1945, Albert Speer attempted to separate General Dornberger from SS authority by appointing him chairman of Arbeitstab Dornberger (Working Staff Dornberger) as part of the Ministry of Munitions. This organization was tasked with breaching Allied air superiority by developing new types of guided rockets for antiaircraft use. However, Dornberger's respite from his nemesis was short lived. Supported by Reichsmarshal Hermann Göring, General Kammler further increased his power by assuming responsibility for the development and production of all long-range antiaircraft rockets and guided missiles. On February 6, 1945, Working Staff Dornberger was transferred to Kammler's own personal technical staff. Thus, the SS ultimately won the battle for control over every aspect of Dornberger's rocket programs.

During the German offensive in the Ardennes, the number of V-2s launched against London had dramatically decreased. In an attempt to hamper the Allies' ability to resupply their forces, the V-2 batteries in Burgsteinfurt and Westerwald, which were targeting the port at Antwerp, were given first priority to the available supplies of liquid oxygen.

On January 28, 1945, the Training and Experimental Battery 444 withdrew from The Hague and traveled to a new testing range, which was planned in the area of Buddenhagen near Peenemünde. A fresh series of improved V-2s were scheduled to be tested from this location. For some unknown reason, the rockets were not ready or available when they arrived, so the unit moved somewhere else temporarily. After January, there is little information in official records concerning the whereabouts and makeup of Battery 444.[4]

In Burgsteinfurt, the third battery of Battalion 485 used the castle estate Schloss Egelborg near the small town of Legden as a firing site in January. By the end of that month, they moved into The Hague to replace Battery 444. During their time in Burgsteinfurt, the third battery of 485 had become quite proficient in firing the V-2s. From the period of November 3, 1944, to January 19, 1945, the unit had fired a total of 153 rockets, with ten being failures.

On January 23, 1945, the second battery of 836 returned from railway experiments in the east. The rail-launched V-2 experiments had been ordered by Kammler in late 1944, even though the whole concept had been abandoned by Dornberger months before. The idea of preparing the rocket under the cover of a double-tracked tunnel and moving it outside just before firing was no longer needed because of the extreme mobility already offered using ordinary road vehicles. The second battery of 836 rejoined the other batteries of the 836 in Westerwald but did not launch until February 28 because of dwindling supplies of liquid oxygen.

Across the English Channel on January 5, 1945, London was subjected to a severe pounding. During a 24-hour period, 21 rockets had been fired by the V-2 batteries in The Hague. The same day, Allied fighter bombers were out hunting targets of opportunity. A railway shipment of B-stoff (alcohol) was attacked between the Dutch cities of Hengelo and Enschede, which resulted in the loss of one rail tanker. Several days later, on January 13, 1945, 22 missiles were launched from The Hague. By the end of January, the average daily number of missiles launched from the coast of Holland had reached its highest point.

Many V-2s were fired from the area of Loosduinen in The Hague. Arthur van't Sant lived with his mother at Thorbecklaan. He was 12 at that time, but he remembers watching the German trucks go by as they pulled the liquid oxygen bowsers along the roads Loosduinseweg to Haagweg in Loosduinen. One of his most stunning memories was when a V-2 malfunctioned and came down near his home on January 1, 1945. The rocket was fired by the first battery of Battalion 485 from Vliegveld Ockenburgh around 5:00 PM. Arthur was looking out through his window and witnessed the rocket rising. Just after leaving the launch table, the rocket turned 160 degrees. It then flipped and flew out of control in Arthur's direction. Arthur hid himself behind a big trunk, hoping it might provide some sort of protection. The V-2 flew horizontally over his house and came down at Indigostraat, on the corner of Kamperfoeliestraat, some two miles from where it was launched. A tremendous boom signaled the destruction of many Dutch houses. Local doctors and nurses arrived very quickly to the crash but were not allowed to enter the area to tend the wounded and killed until after the Germans had collected the remains of the rocket.

On January 16, 1945, the civil authorities in The Hague made a request to the commander of the V-2 troops asking for the discontinuation of missile launches from within the city. They said that "even for the Germans, the crashes caused useless suffering." The German commander, Colonel Hohmann, replied, "All failures were the fault of the Dutch people, because it was only by Dutch sabotage that a V-2 could fail." Curious people who collected the crashed pieces of the V-2 usually had to pay with their lives. The fear of espionage and the smuggling of the rocket secrets prompted the Germans to introduce severe security measures. Several people were executed because they picked up a silly piece of metal from a V-2 crash site.

On January 25, a missile was fired at 8:17 AM from the grounds of Duindigt. It came down at the Archipel area on the corner of Riouxstraat and Bonistraat, totally destroying five houses, with 40 more houses heavily damaged. Ten persons were killed, and 40 more were injured. "We were always listening to hear if the rocket sounded strange, because if something sounded wrong, crashes happened quite often," said Van't Sant.

Throughout the harsh winter, the fighter attacks continued in The Hague. By the end of December, many of the RAF pilots were flying the new "clipped-wing" Spitfire Mark XVIs, employed as a fighter bomber capable of carrying four bombs. A typical day for the Spitfire XVIs started off with an early weather reconnaissance flight over the Channel to the Dutch coastline to determine which targets would be suitable for that day. The squadrons would form up over England and head east toward Holland. As they neared the target areas, the Spitfire pilots would descend to attack altitude, somewhere between 8,000 to 5,000 feet, and in echelon formation they would make their final approach to the target. One method of attack was to fly in echelon over the target until it passed out of sight below the wing, then the Spitfires would roll in and trim the aircraft into a sheer dive. The bombs were released at around 3,000 feet, which gave the pilots enough time to recover from the momentary blackout caused by the g-forces as they pulled out of the dive. They would then be required to make their low-level escape, avoiding the antiaircraft guns in the process. One of the first operations in which the new aircraft were employed was an attack on December 10 against the V-2 personnel located at the Hotel Promenade at Scheveningen. The Mark XVIs laid down their bombs, which exploded directly at either end of the hotel. These days were exhausting for the RAF pilots. During clear weather, a single squadron could mount as many as 40 sorties per day. There were always several squadrons attacking the coast, which resulted in a constant flow of Spitfires over the V-2 sites.[5]

Owing to the many days of inclement weather, it seems the V-2 crews were not particularly hindered by the air attacks. They were able to conduct their daily operations, while taking the usual precautions, on an almost uninterrupted schedule. Even so, the RAF Spitfires of No. 602 Squadron of the Coltishall Wing, along with No. 229 Squadron (which later became No. 603), were the primary nuisance to the rocket troops. The new Spitfire XVIs of No. 602 could carry two 250-pound bombs and extra fuel tankage to operate direct from England; however, once captured airfields on the continent like Ursel, Heimond, and Maldegem became available, the squadrons could fly south to recover and, after being refueled and rearmed, fly again to attack The Hague with 1,000 pounds of bombs. Despite the risk to civilians in the densely populated Dutch city, targets were attacked with noteworthy skill and precision. From January 1945, the primary targets were the storage depots, railway sidings, and road bridges. Many of the attacks in February were mounted against Haagsche Bosch park. During March of 1945, No. 602 Squadron alone completed 469 sorties, dropping more than 1,300 total bombs.[6]

The RAF conducted a major operation in The Hague on February 3 involving 63 Spitfires from No. 453 Australian Squadron, No. 229 Squadron (No. 603), and the No. 602 Squadron along with two photoreconnaissance Mustangs from No. 26 Squadron. Intelligence reports indicated that the tramway sheds near Loosduinen were used by the rocket troops as a filling point for liquid oxygen. The target was believed by the British to be a liquid oxygen manufacturing location, which was incorrect—it was only a parking area for the fueling detachment. From the sky above, the hum of Merlin engines was heard as the Spitfires tipped over side by side and came in to attack. The aircraft made two runs at Langenhorst en Loosduinen: one in the morning and one in the afternoon. Other targets included the Haagsche Bosch and Hook of Holland. On February 21–22, the launching positions of the second battery of 485 were repeatedly attacked by fighter bombers. One of the bombs hit the area of the field store and set it on fire. Allied aircraft had made 1,143 sorties against V-2 targets in the first two months of 1945, dropping 216 tons of bombs.[7]

At the beginning of January, the off-loading of the rockets took place north of The Hague at the city of Leiden, while rocket fuels were delivered at the town of Emmakliniek. Hours before the fuel train would arrive; the tanker trucks and bowser trailers of both battalions would be waiting at the unloading spot.

The railways in Holland were extremely important to the Germans. Not only was it the fastest way to move troops, arms, and supplies, it was also the means by which forced laborers and Jews were moved to the east. The dense Dutch railway network was still largely intact in 1944 and was crucial to the German V-weapons deployment. The Dutch railways were fully integrated into the German lines from the east in order to make the transport of weapons as easy as possible. In September of 1944, under the watchful eyes of the Dutch resistance, the well-organized transports began arriving in the firing areas near The Hague.

The V-2s arrived in the city of Leiden, north of The Hague, at the Leiden main station over the line from Utrecht or by the line from Amsterdam-Haarlem. Leaving Friedrichshaven, a V-2 train took approximately 35 hours to reach Leiden. The main entry points from Germany were Enschede, Oldenzaal, and Winschoten in the north, going through the important transit stations at Deventer, Zutfen, Apeldoorn, Amersfoort, and Utrecht, the main line from the west. Arriving under the cover of darkness, the trains would park under the shelter of the long main station roof. A special detachment of German guards watched over the V-2 railcars at all times. The Group North flak train was positioned near Leiden station with its six heavily armed antiaircraft railcars.

Depending on the arrival time—that night or the following night if it arrived too late—the train would be moved a few kilometers over a connecting line to the cargo off-loading station at Herensingel. Next to Herensingel was the engineer's installation called Pionierspark. Many rockets and rocket parts were always at Pionierspark, as the wreckage from crashed V-2s was also returned to this location. On the siding at Pionierspark, the giant 15-ton-capacity Fries Strabo crane straddled the track while the rockets were lifted from the flatcars and laid onto Vidalwagen road transport trailers. Each train carried 20 rockets, and it took a little over one hour for the engineers to finish the off-loading. Before the light of morning, the road transport vehicles would take the rockets away. The rocket warheads were carried on the same train, but they were not attached to the rockets. They were off-loaded separately and moved to the technical troop's field store to be mounted on the rockets later.[8]

The supplies of A-stoff (liquid oxygen), B-stoff (alcohol), and T-stoff (hydrogen peroxide) were off-loaded from rail tankers under the canvas-covered roof of the Leiden main station. At the Abstellpark (fuel storage area), an average of eight Betriebstoffanhanger (liquid oxygen bowser trailers), four Flugbetriebstoff-Kesselwagen (alcohol tanker trucks), and several other various types of trailers were readied for fueling operations.

The RAF's Second Tactical Air Force and Fighter Command began reconnaissance and strafing attacks on the Dutch railways almost as soon as the rockets began falling on London. Initially, the attacks proved fruitless. But after bombing the vital arteries at Enschede, Almelo, Amersfoort, Utrecht, and Leiden and bridges at several river crossings, the RAF was able to disrupt the transports to a certain extent. Nonetheless, the Germans managed to repair the lines very quickly, even though the damage was at times considerable. On December 11, 1944, No. 602 Squadron RAF dive-bombed the Staatsspoor station, where it was reported that there was another filling station for liquid oxygen bowsers. The pilots witnessed four of the bombs hit on the east end of the station awning, from which a thick cloud of white smoke and fragments rose into the air. Four more bombs fell in the center of the target area among trucks and railway buildings, approximately 100 yards from the station. Other bombs impacted the railway tracks less than 50 yards further to the east.[9]

Allied air power, as mighty as it was, only succeeded in slowing the missile attacks, never bringing them to a halt. If the RAF had mounted a continuing series of air attacks on the Leiden Herensingel station, it probably would have immensely hampered the German efforts and slowed the rate of fire dramatically, but Herensingel was surrounded by civilian housing and therefore was not considered for attack by the RAF. The main railway station at Leiden was attacked by the RAF on several occasions, and during one of these attacks, several alcohol bowers went up in flames in the fuel stor-

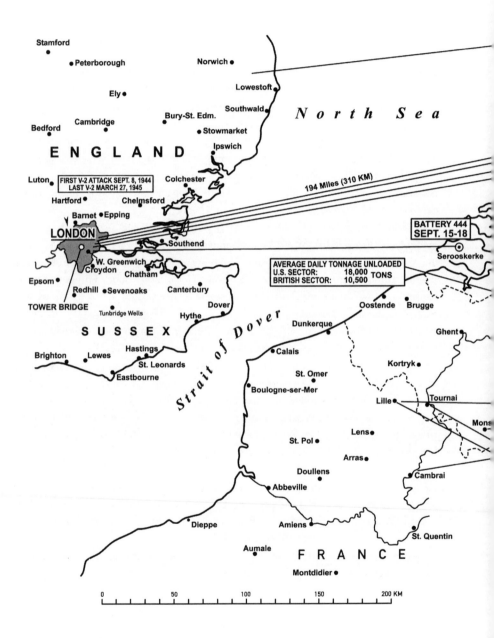

Stamford

Peterborough

Norwich

Lowestoft

North Sea

Ely

Southwald

Bury-St. Edm.

Bedford
Cambridge

Stowmarket

Ipswich

E N G L A N D

Luton

FIRST V-2 ATTACK SEPT. 8, 1944
LAST V-2 MARCH 27, 1945

Colchester

194 Miles (310 KM)

Hartford

Chelmsford

Barnet • Epping

LONDON

BATTERY 444
SEPT. 15-18

Southend

Serooskerke

W. Greenwich

AVERAGE DAILY TONNAGE UNLOADED
U.S. SECTOR: 18,000 TONS
BRITISH SECTOR: 10,500

Croydon
Chatham

Epsom

Redhill • Sevenoaks

Canterbury

Oostende   Brugge

TOWER BRIDGE

Dover

Ghent

Tunbridge Wells

Hythe

Strait of Dover

Dunkerque

Kortryk

S U S S E X

Calais

Brighton   Lewes

Hastings

St. Omer

Tournai

St. Leonards

Boulogne-ser-Mer

Lille

Eastbourne

Mons

Lens

St. Pol

Arras

Doullens

Cambrai

Abbeville

Dieppe

Amiens

St. Quentin

Aumale

F R A N C E

Montdidier

0   50   100   150   200 KM

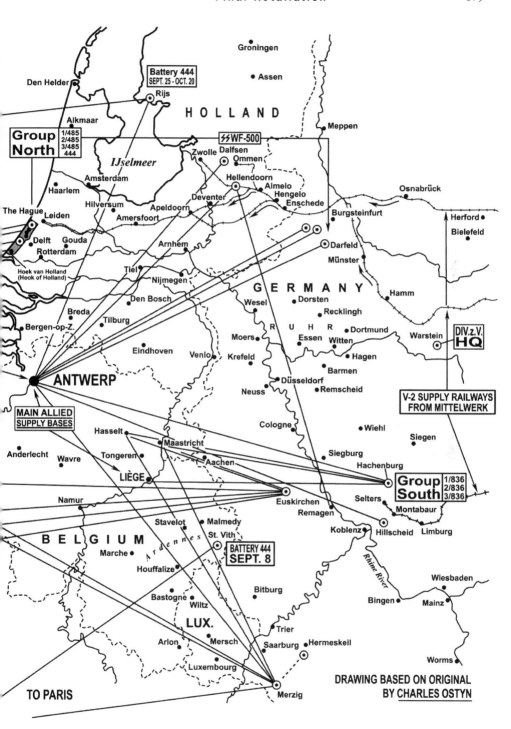

age area. Because of the raids, for a short period in January of 1945, some of the V-2 trains moved on from Leiden through Wassenaar for off-loading at the Hollandsspoor and Staatsspoor stations closer to The Hague.

In the days prior to February 2, 1945, firing sites were reported to be located in the garden of Zorgvliet and in the garden of Stedelijk Museum, where an explosion damaged part of the museum. After February 2, a new firing site was set up at the corner of Statenplein and Willem de Zwijgerlaan. More failures followed. A rocket launched from Klingendael failed, and it impacted near De Battaaf, raining pieces of burning material down on city streets. In the months of January and February, the number of launches substantially increased. Improved transportation of missiles from the Mittelwerk and improved weather in January helped to increase the number of firings, because the firmer soil allowed better missile handling than on the previously wet ground. On February 4, a total of 16 V-2s were fired in less than 24 hours. On the overcast afternoon of February 7, 1945, the first and second firing platoons of the first battery of Battalion 485 moved from the grounds of Ockenburgh near Loosduinen to sites within the Sperrgebiet. The following day, the third firing platoon took up a position at the Hook of Holland.

Further south, at the launching sites in Westerwald, the railway stations involved with V-2 supply in January included a station at Hattert (15 miles west-northwest of Hachenburg) for unloading of alcohol, a station at Hachenburg (on the northern edge of the village) for unloading liquid oxygen and hydrogen peroxide, and a station at Korb (on the eastern side of village) for unloading alcohol; the rockets were unloaded at a station near Erbach (two miles southeast of Hachenburg on the southern side of the town). There was one Strabo crane in Erbach, but another crane was held in reserve nearby. All unloading stations were heavily protected by German flak guns.

As an architecture student in Amsterdam, young Bill Wils was given exemption papers that served him very well. He was allowed to travel freely from Amsterdam to his parents' home in The Hague. In March, he returned to The Hague after receiving word that his parents' home at Bezuidenhoutseweg was heavily damaged during the Allied bombing raid.

"We moved my parents into a rental house that was located on Danckertsstraat. From there I observed the V-2s going up in the air from somewhere not too distant; seemed like perhaps 500 meters or so," said Wils.

The young draftsman made small sketches of the rockets as they were transported along the street in the evening, one after the other, on Vidal trailers hitched to Hanomag SS100s. "After dark, many times the weapon misfired and haphazardly dropped back on the residential areas," he said.[10]

As with the misfires afflicting the Dutch in The Hague, the V-2 launching crews in Westerwald as well as Burgsteinfurt were dropping missiles on their own citizens in German towns close to the firing sites. On January 10, 1945, rocket No. 20950 lifted normally from the firing table at Hillscheid and seemed to be on course. Suddenly, at a distance of about two to three miles high, there was a visible explosion in the tail sec-

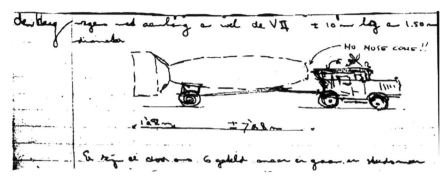

As a teenage boy in The Hague, Bill Wills made this sketch in 1944 showing V-2s on Vidalwagen trailers being pulled near Willem de Zwijgerlaan. (Bill Wills)

tion. The remaining portions of the rocket, including the warhead, fell on the nearby town of Wied. The impact and explosion at the edge of the town resulted in injuries to three local women, along with roof and window damage to houses in Wied and Höchstenbach.

On February 7, 1945, rocket No. 21763 lifted normally, but after 28 seconds, the engine shut off. About 80 seconds later, the detonation was heard. It fell between Hillscheid and Höhr-Grenzhausen in a meadow. The next day, another impact occurred only 150 meters away from the launching table of the third battery at Gehlert. On February 20, a rocket was fired from the concrete pad of the first launching platoon at Hillscheid. It malfunctioned and crashed down on its tail between the firing platforms of the second and third platoons, causing much damage. Between Gehlert and Alpenrod, the projectiles usually flew north past Wahlrod and made an enormous noise as they rose. Local residents saw the rockets ascending in the east with a fiery tail and waited to see if they would crash in the Eifel. The German populace gave the rockets the nickname Eifel Fright. In the middle of March, a V-2 came down in the Stoeffel quarry at Enspel, seriously injuring some of the workers.

At Burgsteinfurt, in the small town of Darfeld, almost every house was damaged in some manner by V-2s during February of 1945. Outside of the small town of Heek in Burgsteinfurt, the second battery of Battalion 485 had been firing at Antwerp on a regular basis from the middle of January to middle of March. Remaining virtually undetected, the crews operated with impunity, as the site was only marginally attacked by "Jabos" on two occasions with no damage. One of the constant problems the crews did face was the wet autumn and winter roads. Special units operated by the SS forced Jewish laborers to continuously repair and reinforce the roads using scrap debris from Allied bombardments. By the end of March 1945, some 627 rounds had been fired from the Burgsteinfurt sites.

In January, the SS 500 had moved to new positions in the area of Dalfsen, about ten miles northeast of Hellendoorn. Near the railway line that ran west from Marienberg to Zwolle, there were new firing sites in a pinewood forest close to the tiny town of Hessum. Many of the local families in Hessum and Dalfsen had to vacate their homes when the SS launching crews arrived. One of these was Willem van Leussen, whose farm was used for the placement of the Leitstrahlstellung (LS) or rocket guide-beam installation. On his property was an apparatus connected to an 83-foot telegraphing dipole mast antenna, which was mounted in the meadows. The apparatus was intended to guide the rocket in the first stage of flight when the rudders could be used in the air or the gas stream to correct the flight path, thereby increasing the accuracy of the rockets by reducing the lateral deflection significantly. The soldiers operating the LS were given their theoretical instruction at Lubmin on the Baltic in October of 1944. During the last days of 1944, the LS troops moved into Holland to join the operations against Antwerp. On December 26, 1944, the beam transmitter was positioned on the farm of van Leussen near Dedemsvaart. The first Leitstrahl-guided shot was launched on January 4, 1945.

The beam transmitter was set up eight to ten miles behind the firing site. It was essential that the beam transmitter be positioned behind the firing location exactly in line with the target. The guide-beam plane was produced by a lobe-switching device with a frequency rate of 60 MHz. A new device was developed, called the integrator or I-device that computed the rocket's velocity without the need for transmitters and was shielded against external interference. Integrated into each of the four fins of the V-2 were various antennas. Protruding from the underside of fins 2 and 4 were pin-shaped antennas related to the Leitstrahl guidance. In about the same location, inside fins 2 and 4, and covered with nonmetallic material were receiving loop antennas for the Brennschluss (engine stop) command.[11]

The use of the guide-beam control during V-2 operations in Holland consisted of several different installations, all working together during the V-2's powered flight. The beam transmitter, code-named Hawaii, was housed in the Senderwagen, which was an Opel Blitz Maultier three-ton truck with a closed cabin. The associated equipment was carried in a second Opel Blitz Maultier called the Gerätewagen. The Gerätewagen carried large cables inside for connecting the antennas to the transmitter truck. The dipole antennas were towed on special trailers. The transmitter truck was set up exactly in the line of fire behind the firing point, and the connecting feeder cables ran out exactly 328 feet on each side at right angles to the two dipole antenna aerials. The duration of the beam transmission was limited to a four-minute window: two minutes of monitoring before launch and two minutes of transmission from one of the seven possible radio frequencies during flight.

The receiver for monitoring the beam position, which had the code name Coblenz, was set up on the same line some 300 to 500 yards behind the firing point. At this position a mast antenna, some 25 meters high, was erected. The receiver would indicate the beam position relative to center zero.

The station that transmitted the signal causing the engine burn to cease, code-named Campania, was about seven miles behind the firing point but could be slightly offset from the firing line. Campania had its own monitoring station, code-named Toscana, situated about 500 yards away.

For surveying the trajectory there were two stations known as Kommando Lt. Lowe, which were located about 500 yards and 1,000 yards, respectively, behind the beam transmitter. The first of these stations was for optical survey and employed a theodolite, while the second station was for electrical survey during operations in inclement weather. There were two other radio stations that were arbitrarily located. One was for communications of the unit, and the other was for checking the quality of the beam transmission. In every case, when the V-2 firing crews moved, so did the wireless troops associated with the position.

The firing sites at Hessum were situated close to the town. A group of four firing platforms were situated close to one another in a coniferous wood to the right of the road from Vilsteren to Dalfsen. All four platforms were on or beside old established roads or paths; the firing points made of cut tree trunks were buried in the soil for firmness. The rockets came straight from the station at Ommen using the road Hammenweg to the field store at Archem. The warheads were attached, and then the projectiles were transported on Vidalwagens via local roads to the launch sites at Hessum. The first V-2 launch from Hessum occurred just after midnight on New Year's Eve. Of the 118 rockets fired at Hessum in the month of January, about 19 malfunctioned and crashed nearby. On the land of local resident Von Martels zu Dankern, an impact crater was made that was 30 yards wide! Because of wet conditions, the launches were moved down the road to the county seat of the Mataram on January 31.

At Mataram, five firing platforms were located in the woods of the elegant estate of Baron Von Vorst. Vehicles were parked near the site, and a number of hard standings with shelter walls were built in addition to those required for the Feuerleitpanzer. In Vilsteren, the technical troops were billeted in a castle with all of its buildings and surrounding grounds, where the work was done under a large half-tubular frame tent on the castle grounds. Nearby the V-2s were transferred from the rail shipments to the road transport trailers. North of the town of Vilsteren, many V-2s were seen resting on Vidalwagen trailers on a sandy pathway near the forest. The area was guarded by German soldiers. While firings were being conducted at Mataram, some of the supply troops seem to have been stationed at the previous launch sites in the forest at Hessum. All German personnel engaged in rocket activities were billeted in nearby houses and villages. A senior NCO was billeted in the large house near the entrance to Mataram park. The LS installation was then situated at Ommerahans, while the unit's flak batteries were positioned behind the cichorei factory in Dalfsen.

The first rocket was launched from the Mataram on February 3, 1945. For a period of more than a week afterward, no more launches took place because of the soggy conditions caused by the thawing of the frozen ground. The heavy trucks and trailers were mired in the roadways. Soon the Germans confiscated people's horses and wagons for the transport of gravel, slag, and wooden planks to firm the pathways for the rockets. Starting again on February 13 and continuing until March 8, the SS crews fired about 80 rockets from the Mataram, with three of the rounds falling close to the firing platforms. In the middle of February, the supply detachment reported that liquid oxygen stores were running low and complained that rail shipments were arriving late and at irregular intervals. When the Kesselwagens did arrive, it was only enough to support the launches for a few days. On the days without fuel, the battery was put through various training exercises or sent out to perform maintenance on equipment or the firing position.

Many high-ranking SS dignitaries visited the SS 500 launching sites, but only SS Panzer General Sepp Dietrich witnessed a V-2 launch up close. SS Gruppenführer Kammler made a special appearance himself on the evening of February 26, 1945, when he visited Dalfsen and the launching sites at the Mataram. The next day one of the misfires fell at the entrance of the dugout for the Feuerleitpanzer. The warhead exploded near the dugout, destroying the control vehicle and causing about 20 injuries.[12]

Heinz Wellmann, firing crew member of the third platoon, remembers his personal experience with V-2 failures: "We could follow the rocket's flight path quite far. Sometimes the rockets exploded in the air or went in the wrong direction. When the V-2 turned itself abnormally in the air, you could tell that something wasn't functioning right. Most went down in the distance with a bang. One time when a rocket lifted from the table, maybe only a meter in height, it fell back. Of course it didn't sit perfectly on the table again, but collapsed. The fuel tanks exploded and the warhead simmered in the burning fuel until finally it exploded, and of course everything nearby was destroyed. You couldn't protect yourself very much. We had our foxholes about 100 feet away, and before each launch one had to crouch down into this hole. We didn't think about our jobs being dangerous beforehand, we always thought it would happen to the other crews, not us. But after we experienced our first failure, we always joked that we hoped 'this thing goes go up' without problems."[13]

In Dalfsen there were many forbidden areas for the local population, yet whenever a convoy of rockets would arrive, the residents in town could see the camouflaged rockets in the street. People were surprised; the SS soldiers made no effort to conceal the rockets, milling about listening to the BBC on their radios. It seemed many of the SS 500 soldiers had grown tired of the war. The strict rules for the Dutch were sometimes relaxed for no apparent reason. Surprisingly, some local residents said the Germans handed out food to their starving children, paid wages to local women for washing their laundry, distributed glass for the shattered windows, and took good care of the occupied houses.

This munificence may have been fostered by a certain amount of integrity pertaining to the battery's senior officer. The soldiers of the SS 500 were endeared to their commander, SS Captain Johannes Miesel. He was considered by the troops to be a respectable and sensible person. Because he cared about the individual solders—their training, their food, and their problems—he garnered the attention of every trooper when he spoke before them at battery meetings. Soldiers felt comfortable stating their grievances to him, knowing he would look into a situation and correct it. Miesel specifically instructed his men to treat civilians and prisoners well. He stated that if the battery ever captured Allied soldiers, they should treat them well—"for my own sons are at the front, and I would hope that nothing will befall them." Later Miesel was transferred to Division z.V. headquarters at Suttrop bei Warstein where he became Deputy Chief of Staff, but before he left, he told the soldiers of SS 500, "If anything should happen to you, you know where you can find me." After Miesel's departure on January 24, SS Major Kurt Tippmann took over as commander of the SS 500.[14]

If a number of German soldiers were amicable toward the civilian population, there was an equal number who were not. Most Twente residents had quite a different view of the German occupation during the war. They remember such things as civilians without proper papers being deported to work camps or SS soldiers shooting on sight anyone who ventured too close to the operational areas. An elderly woman, removed from the Eelerberg grounds during the V-2 firings, was worried about her goats. When she walked to the barn with an armload of hay and a bucket of water, she was shot by a patrol.

Cor Lulof recalls, "You have to understand, the Germans ruled with an absolute terror. My uncle was shot because he picked up a leaflet dropped by the RAF. Around the corner from where I lived, an old lady was shot through the head because she crossed the street ten minutes after curfew. One of the kids in our neighborhood, only eight years old, took a scrap piece of German telephone line to use as a dog leash; he was shot. It was the normal German Wehrmacht who did these sorts of things, not the SS. The SS saw things broader; they murdered the whole male population of some towns. More than 100,000 Dutch civilians, plus 104,000 Dutch Jews, were killed during the war."[15]

The SS 500 left the area of Dalfsen at the beginning of March and returned to the former firing sites at Hellendoorn. Arriving in the Eelerberg on March 7, they began firing again toward Antwerp on March 9. The guide-beam installation was moved from Ommerahans to its final location at Vromshoop in March. At Vromshoop, the beam transmitter was set up with its dipoles on either side of the road leading from the village. The Brennschluss station was situated in the village of Den Ham.

During the war, the western province of Holland was separated from the rest of the country, and there was very little food available during the winter. The Hunger Winter of 1944 was terrible for the Dutch. Children went many times to the Rijswijkse Bos asking the German rocket soldiers for food. In desperation the population began eating things like sugar beets and tulip bulbs. Rita Winter would sneak out at night after curfew and steal vegetables from the neighborhood volks tuintjes (small gardens where people could grow vegetables, fruits, and flowers). On her way she could see the V-2s on trailers rolling down Vredeburgerweg. Because there was no gas or electricity in The Hague, people used anything that would burn for heating and cooking purposes. Behind the dunes, the Germans had fenced off and evacuated a wide part of the land. Civilians tried to get into these empty houses to remove all the wood.[16]

In the spring of 1945 when the hunger in Holland was at its height, the Germans allowed the Allies to fly in and drop food around The Hague. RAF Lancaster bombers, along with American B-17s, flew in so low over the airfield at Ypenburg that Dutch residents could see the faces of the pilots. Swedish ships were allowed to bring in flour to alleviate the severe threat of starvation for the Dutch people of The Hague; one loaf of white bread for each person was distributed.

On March 3, 1945, 61 medium bombers (49 B-25 Mitchell and 12 A-20 Bostons) took off for the Dutch coast. Five of them had to abandon the mission and turn back because of mechanical failures, but the remainder continued on with the mission-not to drop food, but intending to bomb Duindigt and the western portions of the Haagsche Bosch where many V-2s were supposed to be stored. Because the V-2 targets were primarily in wooded areas, Allied commanders believed a low-level medium bomber attack could be carried out without risking too many civilian casualties. However, because of a navigation mistake, the first bombs were dropped to the southeast of the Haagsche Bosch instead of northwest-a deviation of over a mile. The following aircraft also released their bombs on the cue of the lead aircraft.

At that moment, Rita Winter was standing behind a church in a cemetery with her mother and aunt. The funeral for little Frans Winter was concluding. As he was lowered into the ground, the drone of the bombers over Bezuidenhout echoed through the graveyard. The earth under their feet started to shake as the bombs began to fall in the distance. In a short time, the entire Bezuidenhout quarter, the civilian housing south of The Hague forest, was set ablaze. The whole area soon turned dark, as the smoke from the fires blotted out the sun.[17]

Firemen from all over The Hague, Vlaardingen, Schiedam, Rotterdam, and Utrecht en Zaandam battled to extinguish the fires all afternoon and evening. That night, as the fires still blazed, several rockets rose into the smoke-filled skies, launched from Duindigt in the north. The Germans wanted to let the Allies know that they had missed their target once again. Shortly after 2 o'clock in the morning, five firemen were killed when one of these rockets came down at Vlierweg near the Schenkweg, on the edge of the Bezuidenhout. On Sunday the damage was clear: 3,315 houses were burned out entirely, mainly at Thersiastraat and Juliana van Stolberglaan and in the surroundings of Korte Voorhout; 1,217 houses were heavily damaged; 486 civilians were killed; and about 12,000 people lost their homes and all their possessions. What made the event even more tragic was the fact that Fighter Command had used outdated intelligence when planning the raid. The rockets had been moved from the Haagsche Bosch more than a week earlier.

The grounds of the Duindigt Estate were a favorite firing location of the V-2 crews. Located north of the Haagsche Bosch, the easy access given by roads and the many trees in the area provided an ideal location to conduct launching operations. The Germans used this site often from November to February because the area had many different spots where the V-2 crews could operate. When a particular launching platform was damaged by falling bombs or crashing V-2s, they would simply move to another spot, sometimes only yards away. In total there were about 30 launching sites at Duindigt, either on the grounds of the estate or the nearby horse racing track. The supply of rockets was very close with the field store located in the woods of Het Ravelijn just to the north in the woods off the dune. On February 13, several firing sites were situated close to the grandstands of the racetrack, straight across Wittenburgerweg on the track and behind a house at Buurtweg. The Dutch resistance had reported that trucks and vehicles for the rocket troops were parked at Rust en Vreugdlaan, at the entrance Houtlaan, and also at the Buurtweg. The greatest V-2 activity here was in mid-February. Inside of one week, over 70 V-2s were fired at London, with as many as 50 percent of them hitting greater London.

In early March, RAF Spitfires bombed the Duindigt launching sites incessantly. After several days of constant bombardment, there was no place on the estate that remained undamaged. The cratered roads were

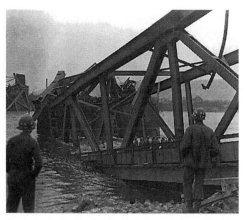

The Ludendorff railway bridge at Remagen collapsed into the Rhine River ten days after its capture by American forces. Hitler personally ordered V-2 missile strikes on the bridge. On March 17, 1945, the SS Werfer Battery 500, stationed at Hellendoorn in Holland, launched 11 rockets towards the bridgehead. One rocket impacted less than 300 yards from the bridge, shaking the ground tremendously. Following several more impacts in the area, the already weakened bridge collapsed into the river around 3:00 PM. (*NARA*)

impassable, and the Duindigt area was never used again by the V-2 crews. A few days later, some Dutch people came to view the estate, and they were overwhelmed. The once-beautiful estate did not exist anymore. They were shocked to see the 200-year-old trees that had been blown apart. The two large manors, five smaller houses, and the farm were in ruins. Even larger than the bomb craters were the immense craters dug by malfunctioning rockets on the estate grounds. On March 17, a new firing site was positioned between the villas on the estate of Groot Hazebroek, but these didn't last long. The final days of the campaign in The Hague became ones of abandonment, with a last few spiteful missiles being fired from open roadways.

The last rocket to cause great loss of life in Central London came down at 11:10 AM on March 8, 1945, directly in the middle of the lively Smithfield Market at Harts Corner in Farringdon Street and Charterhouse Street. The rocket went through the base of the building and into the railway tunnels beneath. As the floor caved in, many people inside were consumed when structure above collapsed into the hole. Others were killed on the exterior of the buildings. One hundred sixty persons were killed and 123 seriously injured.

In early February, the many vehicles belonging to the second battery of Battalion 836 were loaded on to railcars in Tuchel to be transported once again to the firing front at Westerwald. The soldiers of the battery had been away for some months in Tuchel and Peenemünde conducting experimental shootings. After traveling for many hours, they were off-loaded near Aumenau. From the railhead they drove their vehicles over the country roads through Grävenwiebach, Seelbach, and then Falkenbach, finally arriving in Wirbelau. To the west, above of the town, the streets were closed from the crossing at Wirbelau to the crossing at Eschenau, and the complete forest area was secured. Everywhere were armed soldiers and signs reading Restricted Area. Off the main road, about 500 yards into the forest, there was a freshly cut network of vehicle paths, along with three stationary launching platforms (concrete pads), constructed by the engineering platoons. The three firing platoons manned these new positions and waited on the delivery of the rockets while the technical and supply troops took up positions on the sports field in Wirbelau. Finally a few rockets were delivered on Meillerwagens. It had been difficult for the long trailers to pass through the small streets of Wirbelau. The rockets were lifted up into position on the firing tables, waiting on fuel. The crews had lots of spare time, because the fuel never arrived. Firing crews began to notice shortages of supplies in February of 1945. The rockets and fuels were still delivered, but on a much more inconsistent basis.

The construction of the new positions at Wirbelau proved to be a logistical error. The available supplies of liquid oxygen were growing increasingly sparse, and Wirbelau was too far away. There would be too much evaporation (from the time the fuel was transferred at the railhead to the time it was delivered to the firing sites). At the end of February, the second battery of 836 was again loaded onto a train at Aumenau. Moving through Limburg, they arrived at Kirburg. This was firing position No. 313. These new emplacements were close enough to the Gehlert sites so that both firing areas could share the same Brennschluss signal. From February 28 to March 16, 19 V-2s were fired from the Kirburg against Antwerp. The vicinity was bombed and strafed nearly every day, several times a day, by U.S. fighter bombers. On some occasions, the V-2 crews launched rockets in plain view of the P-47 pilots of the Ninth U.S. Army Air Force, who had been circling Kirburg. On March 16, the third battery of 836 fired four rockets from sites at Gehlert. That same day, the second battery launched a double shot followed by third shot just a few minutes later from Kirburg. These were the last rockets launched from the Westerwald area. Because of the failing fuel supply and the Allied advance at Remagen, Battalion 836 became the first V-2 launching unit to permanently halt operations.

On March 19, 1945, transportation was arranged for the second battery of 836 to move out of their positions. The movement was to be carried out during the night. From Kirburg to Marienberg, the unit drove to Lüneburger Heide. Soon plans were made for the rest of the unit to leave the area. During the movement of the engineering platoons from Langenbach into the area of Marburg, the truck column was attacked and damaged and left stationary on the road the whole day. It was repeatedly attacked by Allied aircraft and totally destroyed. The first battery of 836 was not able to load their equipment for evacuation, as the marshaling yard at Firckhofen was badly damaged by P-47s and no trains could come in to get them. In Hachenburg, surplus V-2s and supplies were loaded onto a long train. Another train (this one with excess alcohol) was connected to the Hachenburg train. It took some time before the trains were underway because all available locomotives were in need of repair. The American bridgehead at Remagen was increasing but had not yet reached the rocket troops. Soon the units used any means available to leave the Westerwald area.

March 7, 1945, was a drizzly, gray day in the small German town of Remagen. Wehrmacht troops were guarding the old Ludendorff railway bridge across the Rhine River and had already prepared the bridge for demolition. American troops of the U.S. First Army surprised the German garrison and moved quickly to seize the bridge. With American infantrymen and tanks gathering on the opposite side of the bridge, the German commander finally decided that the bridge should be blown. However, a problem arose when an American artillery shell severed the lines to the main charges. The German defenders could only manage to detonate a small charge, which raised the bridge in the air. However, to the German's surprise, it settled back onto its foundation, seemingly intact. There was structural damage to one of the side supports on the far tier, but the bridge was only weakened. American infantry fought their way across the bridge and secured it later that evening. The capture of this strategic link across the Rhine was a tremendous victory for the Allies and a devastating development for the German defenders of the Ruhr.

Over the next ten days, the Americans poured thousands of troops and supplies over the Rhine and into Germany. Hitler ordered the immediate destruction of the bridge using any and all means possible. The Ninth and Eleventh Panzer divisions hurried to engage the American Ninth Division, the objective being to push the Americans back across the Rhine and destroy the bridge. On March 8, ten Luftwaffe aircraft, including eight Ju-88 Stuka dive-bombers, attacked the bridge, scoring two hits. On March 15, 21 German jet aircraft attacked the bridge with poor results. U.S. antiaircraft batteries shot down more than half of these fast-moving aircraft.

Hitler then ordered the use of other weapons against the bridge. These included a huge 17-centimeter railroad artillery gun, intrepid underwater scuba men (they were to swim down the Rhine in an attempt to place demolitions on the piers), and finally V-2 rockets to strike the bridgehead area. Even though the bridge was probably the aiming point for the V-2 attacks, Hitler's objective was not to strike the bridge so much as it was to disrupt the whole area with the rocket attacks. He envisioned 50–100 missiles over a two- to three-day period, but at this point in the war there was no way the rocket troops could muster such an effort.[18]

On March 14, 1945, General Bayerlein received an urgent communiqué from the Führer concerning the Allied bridgehead at Remagen. The General Staff was shocked when they learned Hitler was ordering the use of V-2s against targets on German soil. When the Führer was reminded of German civilians living in Remagen and the surrounding areas, he demanded the rocket attacks go forward regardless of casualties to civilians. Late that evening, because of the accuracy problems with the V-2, the German soldiers fighting in the bridgehead were pulled back about nine miles from the front lines. This move puzzled some of the Allied commanders at first. However, contemplation of a coming V-weapon attack on the bridgehead had been discussed.

The U.S. commanders decided to keep any talk of this possibility hushed and not spread among the troops.[19]

General Kammler received orders to attack Remagen and immediately issued commands to the SS 500 crews at Hellendoorn, 130 miles to the north of Remagen, to fire on the American bridgehead. The SS 500 had been firing since March 8 from the older sites at Hellendoorn, 130 miles north of Remagen. Despite fuel supply problems, the troops were able to muster 11 rockets for the attack on March 17, 1945. If the unit was given the assignment because of their employment of the guide-beam equipment, the decision was erroneous. The Leitstrahl apparatus could not be used for the attacks on Remagen because the installation was already situated for semi-permanent operations in the direction of Antwerp; it could not be moved on such short notice. Nonetheless, the War Diary of the SS-500 reveals that the crews began firing at 9:45 AM and continued throughout the day until the eleventh round tilted toward Remagen around 9:45 PM. All impacted near Remagen except for one, which came down some 40 kilometers away near Cologne.[20]

The first round airburst over Kasbach with portions falling within a 1 kilometer radius of the bridge. The warhead most probably impacted closest to the bridge, coming down at 9:50 AM in the backyard of a house belonging to a farmer named Herman Joseph Lange. A dozen U.S. soldiers were billeted in this house; three of them were killed instantly. The rocket impacted less than 300 yards from the bridge shaking the ground with a tremendous boom and mini-earthquake. In the explosion, 18 farm animals vanished. Around 12:20 PM, another V-2 fell about a half a mile from the bridge, close to the Apollinaris Church in Remagen. This round destroyed several buildings near the church including the command post of the 284th Combat Engineers Battalion, killing another three men and injuring 31 others and causing collateral damage to buildings within 1,000 meters. The impact shook every structure in the city. At 3:17 PM there was a huge detonation 300 yards north of the parish church of Nierendorf. At 3:44 PM, a V-2 came down about 3.5 miles northeast of the bridge at Oedingen killing six people and destroying many houses. Minutes later, another fell directly into the Rhine River a little less than one mile from the bridge. Around 6:15 PM another rocket broke up in the air about one mile southwest of Heimersheim, which was followed by another round that impacted northwest of Strödt, near Koblenz.[21]

Even before the last of the 11 rockets had been launched, General Kammler sent a request to Army Group B for a report on explosions observed in Remagen bridgehead. He wanted details of place, time and effect. Because of the size of the target area and the features of the terrain, accurate spotting from troops and artillery observation posts was extremely difficult—only one report was given. The Volks Artillery Corps 409 described some strikingly loud noises of explosions and reported a powerful detonation at about 9:50 AM near the river.[22]

Immediately after the V-2 attacks of that morning, the already weary Ludendorff Bridge fell into the Rhine River sometime after 3:00 PM. If the earth-shaking rocket attacks did not substantially contribute to the collapse of the bridge, it is certainly a

V-1 damage at Montignystraat in Antwerp on March 27, 1945. Later the same day the final V-2 to impact Antwerp, came down at Antwerpsestraat / Lieven Gevaetstraat, Mortsel, causing 23 casualties. (Dr. Kenneth Hartman)

strange coincidence that the bridge fell into the river the same day. In describing the 1.4 billion foot-pounds of force created by a single V-2 hitting the earth at 3,000 mph, General Dornberger touted it equal to "fifty locomotives, each weighing a hundred tons, impacting the ground at 60 mph." However, the majority of the V-2's energy is converted to heat on impact—much of the shockwave goes up and in the soil much more energy is converted to frictional heat. In a dense city the blast wave from the speed of impact could cause considerable damage to nearby buildings, but steel bridges are generally very difficult targets to bring down. During WWII, there are almost no examples of bridges destroyed by anything else than a direct hit or demolition charges. Even the atomic blasts in Hiroshima and Nagasaki had amazingly little effect on bridges. Considering the constant shelling over the past days, the tons of equipment crossing the river, and the enormous weight of wooden planking added to the bridge, the force exerted by the V-2s crashing in the bridgehead simply added to overall stress on the already weakened bridge. The following day, even though he could have only assumed the rocket attacks had been the catalyst, Hitler sent his personal congratulations to the crews at Hellendoorn for the destruction of the Ludendorff Bridge. The firing crews boasted of their accomplishment to Dutch civilians the next day.[23]

At the conclusion of the RAF anti-V-2 campaign in The Hague, it was difficult to judge the overall effectiveness of the operations. Even during the last days of operations, Allied airpower found it virtually impossible to locate and attack the actual V-2 firing sites. Tactically, it was found that little damage was inflicted by Allied air attacks on the launch sites. The attacks caused few casualties among the rocket personnel; the greatest material damage to the launch sites was actually caused by malfunctioning rockets exploding on the platforms. Recognition of the launching points from the air proved to be extraordinarily difficult. Even though pilots witnessed V-2s launched directly in front of them, there were no occasions when launches were unmistakably prevented. It seemed to be impossible for pilots to correlate rockets rising from the forest with a particular launching position. Numerous launching sites were never discovered. British radar had tracked the launch of all 11 rockets fired at Remagen on March 17, and on several occasions, Dutch resistance organizations in Overijssel had reported the exact launching positions of the SS 500 to Allied officials. The accumulative data gave them a fairly good idea of where the firing sites were located. Nonetheless, the sites at Hellendoorn were never attacked. On the other hand, during some periods, the numbers of rockets launched decreased, particularly on the Dutch coast, and it may well have been because of the activities of the Spitfire XVIs. At the very least, the air attacks demonstrated to the people of The Hague hope for the eventual German defeat.[24]

As the Russian Army closed in on Peenemünde from the east in 1945, it became apparent to von Braun and his staff that things were coming to an end at Peenemünde. Von Braun had received several contradictory orders from German command, which was in mass confusion at the time. As von Braun later stated, "I had ten orders on my desk. Five promised death by firing squad if we moved, and five said I'd be shot if we didn't move." Von Braun's staff was under the direct command of the SS General Kammler. It was feared the Kammler might use the scientists as a bargaining chip or have the scientists killed just to keep them from being captured by the Allies. Since he was damned either way, von Braun called a meeting very early on in mid-January with the remaining top officials at Peenemünde. The Red Army was approaching from the south, and the path of escape might be closed soon. If the scientists and engineers remained at Peenemünde, they would either be killed in combat or taken prisoner by the Russians. They certainly wanted neither. They all decided that if it were possible, they would surrender to the Americans. If nothing else, they wouldn't be killed, but they reasoned it was more likely they could continue their research under American supervision after the war. They knew that somehow they needed to smuggle all their research papers and important equipment out of Peenemünde. They certainly could not allow a decade's worth of work to be destroyed or fall into Kammler's hands.

There was no chance Peenemünde could be defended. That is why von Braun was utterly amazed when he received an order from the local army defense commander to fight the Russians when they arrived at Peenemünde. This would almost guarantee their demise. But another set of orders came from Kammler stating that the engineers and scientists were to move to central Germany, close to the Mittelwerk factory. Von Braun was still wary of Kammler's real intentions. Kammler might be moving the scientists to a location where he would have the ability to turn them, along with the technology at the underground Mittelwerk, into hostages. Nevertheless, he figured Kammler's directive was the best option for their continued freedom.

Von Braun prepared to evacuate thousands of engineers, scientists, and their families to central Germany. It was a tremendous task, but he insisted that it be done in an orderly fashion. He was Peenemünde's accomplished leader at this time. For ten years he had showed his leadership abilities with staff and technical problems and in dealing with politicians, but this move to the south truly showed power of von Braun's determination. German command and society was crumbling all around them, yet somehow his organization held together.

They went to work rapidly. Almost all of the coordination went through von Braun's close staff. Simple things such as procuring boxes became a daunting task at this point in the war. They invented a color-coding system to make it easier to identify the contents of what they were moving. A convoy was organized, in which thousands of workers, engineers, and other Peenemünde residents would be transported by train, truck, car, and any method available. Moving this many people was bound to draw attention, and von Braun knew he would be questioned about the move by local authorities. As luck would have it, a recent shipment of stationary from the SS, which identified Peenemünde personnel as a branch of the SS, was badly mangled at the printer. The letterhead was supposed to read BZBV Heer, the name of an organization within the SS. Instead, it read VABV, the initials of a nonexistent organization. Von Braun's staff quickly invented a top-secret agency with the initials VABV, translated in English as "Project for Special Dispositions."

The initials VABV were painted and marked on boxes, vehicles, armbands, and anything that might be checked by SS inspectors or other authorities. All of the material and equipment was then packed into trucks and cars. The convoy headed south, and along the way SS agents stopped the caravan frequently, but the VABV trick worked, and they were allowed to continue.

On February 27, 1945, von Braun and his driver were leaving Peenemünde for the last time, speeding through the mountainous terrain, when they both fell asleep. The vehicle plunged down the cliff side, killing the driver. Von Braun suffered a broken arm and fractured shoulder. He awoke to find himself in a hospital bed. Even though he was in no condition to be up and moving around, von Braun insisted that his arm be set in a cast, and he was back to supervising the convoy. They arrived at their first destination at Bleicherode. Soon they received word that Peenemünde had been captured by the Russians. General Kammler ordered von Braun and 500 of the top scientists to be separated from their families and moved to the village of Oberammergau. They were placed in a small internment camp that was, in von Braun's words, "extremely plush, notwithstanding the barbed-wire around it." Kammler was indeed holding the scientists hostage. They were surrounded by SS guards constantly. One day von Braun pointed out to the head of the SS guard that the Oberammergau camp could be easily bombed by Allied aircraft. One attack could wipe out all of the Third Reich's top rocket scientists. Any guard that allowed that to happen would surely be shot. The guard agreed and let the scientists out of the camp and into the streets of Oberammergau. He also agreed to let the scientists dress in civilian clothing so

American troops would not suspect that they were of any importance. Von Braun quickly arranged for vehicles from Bleicherode to come get the scientists. They were really free at this point. Now all they had to do was wait and then surrender to the approaching Americans.[25]

On the evening of March 23, 1945, British troops under Montgomery had reached the Rhine River, and several days later, the situation for the Germans in Holland became critical. The German High Command ordered the immediate withdrawal of all troops and material related to both the V-1 and V-2 operations in Holland. The order stated that it was imperative that all expensive material, along with the specially trained troops, return to Germany before they could be captured by the advancing enemy. The final six rockets were fired at London on March 27, including one that destroyed a block of flats at Whitechapel in Stepney. One hundred thirty-four people were killed and 49 others were injured. It was a bitter end to the rocket misery that held its grip on London for so many months.

In The Hague, the rocket stores of Battalion 485 still housed about 60 rockets. Every rocket or piece of equipment that could not be transported by truck or tractor was loaded onto a long cargo train. At an assembly site near the centuries-old Kasteel Duivenvoorde, various scrap rocket parts not worth transporting were heaped into a large pile south of Rijksstraatweg and wired for demolition. On March 29, at around 2 o'clock in the afternoon, an enormous explosion shook the thick walls of the old castle. Windows of houses in a wide area were blown out, ceilings and walls came down or were torn apart, and trees were snapped like twigs into many pieces. The rocket parts that survived the explosion were thrown into the canal waters at Leidschendam. This was the farewell of Battalion 485. Protected by low clouds on the rainy Thursday afternoon, they drove in a long procession to Leiden, from there to Utrecht, and then finally to the German border. Equipment removed from the firing sites was littered along the route as fuel ran out and chain of command broke down. They arrived in the area of Fallingbostel on April 1. They were joined there the following day by the second battery, which had fired the last two operational V-2s of the war on the evening of March 28, 1945, from the Strönfeld site near Heek in Burgsteinfurt, the last rocket falling in Holland about 11 miles northwest of the Antwerp harbor in the neighborhood of Ossendrecht. A few days later, the majority of the battalion's specialized vehicles and equipment, along with some remaining rockets, were destroyed haphazardly with explosives at a large outdoor storage area for V-2s near Leese, northwest of Hannover in the Hahnenberg Forest.[26]

Since September of 1944, the rocket crews on the Dutch coast had aimed approximately 1,300 rockets at Britain. Of that total, more than 500 had fallen in the London Civil Defense Region, around 500 more elsewhere on land, and 50 to 60 just offshore. The remaining were rockets that had malfunctioned in some manner. The official overview said 2,724 people lost their lives and 6,467 people were severely injured.[27]

The last V-2 to strike greater Antwerp landed in Mortsel on March 27, 1945, killing 27 people and injuring another 62. The final flying bomb attacks occurred on

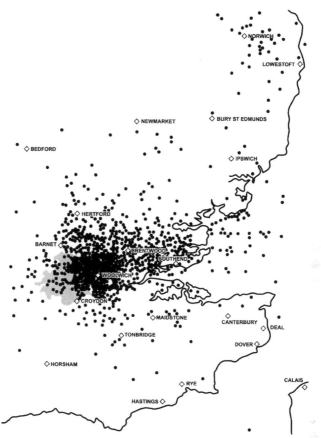

Plot of the 1,115 reported V-2 hits in England, 517 of which struck greater London.

March 30. The SS 500 had been firing on Antwerp from sites near Hellendoorn through the period of March 9–27. During the last days of the month, some of the platoons were given a quick course in close-combat infantry training. Being completely cut off from outside news, the soldiers could only guess as to why this training was given. On the nights of March 27–30, the whole battery packed up and withdrew from Hellendoorn because of advancing Canadian forces. Each platoon left the launching area on different nights using different routes, but all went in the general direction of Nordrhein-Westfalen. They took everything, including excess rockets and all battery vehicles, moving under the cover of darkness. As the units traveled through Lingen, Niederdorf, Lemke, and Nienburg, the firing crews thought it was a normal firing position change. The soldiers would only find out days later that the SS 500 had been ordered to move east in defense of Berlin. On April 1, some of the platoons arrived at Dorfmark just north of Fallingbostel.[28]

In the area of Fallingbostel, 23 miles north of Hannover, the men of the SS 500 were split up into new groups. Some received new combat assignments, while most formed the nucleus of several new Nebelwerfer artillery batteries. The 15-centimeter Nebelwerfer 41, or Screaming Mimi as American soldiers called it, was an artillery piece that fired solid-fuel rocket projectiles. The weapon was designed to saturate a target with spin-stabilized smoke, explosive, or gas rockets. A large portion of SS 500 soldiers chosen for the new batteries was previously trained as artillery men prior to their V-2 deployment. The artillery pieces were brought in from the Nebelwerferschule (Nebelwerfer School) at Celle, and a quick training program was instigated by elements of the SS Werfer-Lehrabteilung. On April 5, most of the specialized V-2 equipment and vehicles belonging to the SS 500 were demolished to prevent their capture by the Allies. However, the Feuerleitpanzer halftracks belonging to the platoons were converted into combat vehicles by cutting the armored top section away and fitting them with twin 20-millimeter guns.

There is some information that seems to indicate that Training and Experimental Battery 444 fired additional rockets from the area around Verden in Germany as late as April 5. The unit had reportedly already fired two new test series V-2s in the direction of the North Sea from this area a few weeks earlier, sometime in the first days of March at Armsen (southeast of Verden an der Aller). Eyewitness reports said that on April 5–6, there were two V-2s launched in the direction of the North Sea from Lüneburg Heath, with one malfunctioning and crashing nearby. Days later, the unit drove to the village Steinhorst where a portion of their equipment was demolished. Some of the men formed up with orders to join the V-2 division and fight the Allies as infantry, but a larger part of the men were tired of the war and deserted to the north. These men arrived at a place called Welmbüttel in Schleswig-Holstein. The remaining rockets and specialized vehicles, such as the Meillerwagen, were towed into the bog at Welmbüttel and demolished with explosives. Some of the soldiers took quarter in nearby villages and reportedly stayed there until the war ended.[29]

Battalion 836 had been idle for several weeks. Having launched their last rocket from the Hachenburg area on March 16, 1945, they spent the last days of the month packing and shipping excess rockets and equipment out of Hachenburg because of the lack of liquid oxygen and the Allied breakthrough at Remagen. They received word to gather at Bramsche, about ten miles west of Osnabruck, for further orders. The deterioration of the basic military situation, however, prevented this. Instead, the battalion was ordered to the area of Visselhövede for the Blucher Undertaking. From there, their remaining rockets were to be fired against the Küstrin Fortress in Poland. Over the previous two weeks, Stalin's forces had advanced roughly 100 miles from the Baltic, near Kolberg in Pomerania, to the Oder fortress of Küstrin, about 60 miles northeast of Berlin. The plan fell apart because of the total breakdown in Germany. On April 3, 1945, Hitler ordered that no more explosives were to be used for V-2 warheads, thus terminating the offensive once and for all. As a result, all of Group South's equipment was destroyed on April 7, 1945, in the area of Celle to prevent its capture by the Allies.

The individual soldiers of the rocket batteries believed their V-2 mission to be an important one. They were disheartened by the rapid retreat and sudden change in their situation. The war diary of the Battalion 836 stated on April 8, "With all of our specialized equipment destroyed, the long-range rocket group has lost its character as an elite unit. Time is up for Gruppe Süd and the employment of the V-2. We are now nothing more than an ordinary infantry combat group."[30]

On April 6, 1945, as the British forces approached the Weser River, chaos was everywhere. The German units were disorganized and communications were nonexistent. Near Stolzenau some of the excess SS 500 personnel, along with a battalion of Hitler Youth and a company of inexperienced German engineers, were ordered to fend off enemy tanks with only small-arms fire and a few 88-millimeter flak guns. Tanks of the British Second Army attempting to cross the river were eventually turned back but crossed later at Petershagen, south of Stolzenau. The tanks headed north to meet up with a British commando brigade near Leese, where on April 8, abandoned and demolished V-2s were discovered on flatbed railcars and in the forest just outside a chemical manufacturing plant. Kampfstoffabrik Leese was one of a number of secret plants built in this area of to produce chemicals for the German war effort. Just to the north was another company, Riedel de Haën AG, which produced teargas, hydrogen peroxide, and sodium permanganate. On April 10, British forces discovered several V-2s on railway wagons when they captured a large munitions factory eight miles southwest of Nienburg near Liebenau. As the Allied push from the west drew closer, the leftover remnants of the SS 500 began a disorderly retreat to the area near Göhrde on the west side of the Elbe. In the following days, Battalion 836 moved to the marshaling areas east of the Elbe River where they were joined with Battalion 485 and a portion of Battery 444 near the area of Dannenberg. The V-2 division was now at the disposal of the German 41 Army Corps under Lieutenant General Rudolf Holste.[31]

The tiny town of Bromskirchen, located northeast of Westerwald, about 15 miles north of Biedenkopf, was captured by Combat Command B of the American Third Armored Division on March 29, 1945. In the morning, a train was coming into Bromskirchen from the marshaling yard at Allendorf-Eder. The train had many flatcars and was exceedingly long with hardly an end in sight. At the head of the line was an oil-fired locomotive, and further back was a coal-fired locomotive. After moving a few miles, it so happened that the train came to a stop near Bromskirchen because the coal-powered locomotive ran out of water. The operators of the train did not want to be stuck there in the open for fear of being attacked by the dreaded Jabos. Quickly they disconnected the train in the middle, and the commanding officer ordered the first half of the train to continue under the power of the oil locomotive toward Winterberg. For some reason, this portion of the train only made it to the tunnel at the Brilon Forest. When it was discovered by the Americans later that day, they found seven railway cars, each containing 12 V-2 warheads, one car containing boxes of carbon-graphite V-2 rudders, fuses, batteries, and cans of calcium permanganate.

The remaining section of the train with the steam locomotive waited about 20 minutes at Bromskirchen to fill up with water. Suddenly an American tank appeared on the road and opened fire on the train with its machine guns. The locomotive received a direct hit, damaging the main steam pipes and leaving a large hole in the side of the locomotive. The railway workers jumped from the train and escaped. After its capture, the Americans found ten railway cars containing nine damaged and partially burned V-2s, plus the scattered parts of another V-2 and some warheads. Elements of the Third Armored Division also captured a factory nearby located at Hatzfeld, which contained several intact V-2 rockets.

With the discovery of these rockets, completely intact, members of the western press got their first chance to see the infamous German V-weapon up close. The discovery was in the limelight for several days. Even Supreme Allied Commander General Eisenhower traveled to the zone to scrutinize the curious find, as English and American newsreels reported his inspection in detail; it was a sensational story. The missiles were then loaded onto large trucks and taken to Antwerp; from there they were shipped to America.

On April 10, tanks and infantry of the U.S. 83rd Armored Reconnaissance Battalion overran German rear guard elements west of the Mittelwerk near Bleicherode, dueling with several German Panther tanks in the area. One of the platoons cleared the V-2 refurbishing plant at Kleinbodungen and found a number of complete V-2 rockets, minus warheads, lying about on jigs.

Since February of 1945, there had been a plan by Kammler to convert the V-2 launching units into a normal motorized combat division. Many officers in the division thought this would be a senseless waste of human lives. Many recognized the final defeat for Germany was inevitable and saw the only sensible solution as being an organized surrender. On March 10, several division officers, including Colonel Thom, conducted a meeting at Bad-Essen. They all agreed the division should surrender and only sympathetic colleagues should be informed of the plan. Colonel Thom, Chief of Staff for Division z.V., blocked Kammler's ordered conversion by simply delaying the reorganization of troops for the required battle groups.

It was understood among all that the idea of surrender should not be mentioned to SS General Kammler. It would have been suicide to do so. Kammler's fanatical attitude could not be reasoned with, and any suggestion of surrender would bring an immediate death sentence. Kammler had already given the formal order to reorganize the rocket troops into infantry regiments, and a few weeks afterward, he made a speech before the regimental and battalion commanders, giving wild battle orders. Soon after that, Kammler disappeared from Division z.V. affairs altogether. Colonel Thom reasoned that if the division surrendered as a whole, the Allies might be interested in employing some of the highly trained V-2 specialists with their practical field experience for development of rockets of their own. He believed, correctly, the Americans and the British would not want the German rocket troops to fall into Russian hands. Thom was then called back to Berlin at the end of March. It is unknown if his confidence was betrayed after he sought support from higher echelons in Berlin, but for whatever reason, he was relieved of his post. It looked as if the surrender to the Western Allies would not happen.

By the middle of April, the V-2 division, under command of the 41 Army Corps, crossed to the east side of the Elbe River near the town of Dömitz. As part of a motorized infantry division, they were assembled in the area of Lenzen, ready for deployment against the Russians. On April 26, 1945, the acting commander of the division, Lieutenant Colonel Schulz, former commanding officer of Group North, was persuaded by SS Lieutenant Colonel Wolfgang Wetzling to sign an order issuing authority to open surrender talks with the Americans. Before this could occur, once again the division found itself with a new commander when Schulz was relieved by the young Colonel von Gaudecker.

Colonel Wetzling did not give up. He discussed the surrender plan with several staff officers of the division. One of them, Major Matheis, agreed to go with Wetzling to see the new commander of the division. During a short meeting on April 29, Colonel von Gaudecker agreed to the surrender. The plan was discussed in detail with the wishes of the division written down. Wetzling and Matheis would act as emissaries accompanied by a signals officer, and the arrangements would be transmitted back from the Americans on a special wireless wavelength. On the night of April 29–30, von Gaudecker gave two signed papers, one in German and one in English, to Wetzling and Matheis. They received another order for fictitious duty in the area of Lenzen in case they were stopped by German security patrols. The three men started off into the darkness to find the Americans.

The next afternoon, they reached the banks of the Elbe River, close to the small village of Wootz. Despite carrying a large white flag, they were met with mortar fire. Major Matheis was slightly injured by shrapnel. After continued shouting and waving of the flag, they finally got the attention of some American soldiers on the other side. Several hours later, the Americans transported the emissaries across the river and brought them to the local American headquarters. Later they were blindfolded and taken to another headquarters. There, talks began between the German emissaries and 15 officers of the U.S. 29th Infantry Division. American Colonel McDaniel presided over the talks and told the Germans he was authorized to receive their declarations.

Colonel Wetzling announced the readiness of the V-2 division to surrender en masse, unconditionally, provided an assurance was given that members of the division would not be handed over against their will to the Russians. McDaniel replied that the unconditional surrender of the entire V-2 troops was accepted and that the members of the division would be treated according to the Geneva Conventions. He was, however, not entitled to give further assurances. The emissaries further declared that the V-2 division would also bring with their surrender the goodwill to cooperate in the further development of the rocket for the good of the Western Allies. Colonel McDaniel agreed that the surrender of the V-2 division would be kept secret to prevent reprisals against the families of the division's officers. The declaration was accepted by the Americans and forwarded on to the higher American authorities.

Colonel McDaniel said the division should cross at the same point near Wootz. Boats of all kinds, including DUKWs (amphibious trucks), would be in position by the morning of May 1 to ferry the division across the Elbe. It was not known how long it would take for the V-2 division to arrive, but the Germans were instructed to assemble near Lenzen and to move from there in a line with every tenth vehicle carrying a white flag. At the crossing point, four white flags were to be erected 50 meters apart from each other. The Germans were assured that their movements would be safe from American fighter aircraft. Colonel McDaniel asked for all V-2 equipment, special vehicles, and V-2 documents of any kind to be handed over to the Americans. The Germans stated that all equipment of this kind was destroyed when the V-2 troops retreated from the operational launching sites.

Without any sleep, Wetzling and Matheis returned to their division headquarters. They arrived late in the morning to find new difficulties had arisen. The situation had deteriorated considerably because of Russian advances nearby. Colonel von Gaudecker and Major Schuetze were in the middle of giving tactical commands to units battling Russian forces. Von Gaudecker said it was impossible at that time to withdraw the division from its positions. The situation became even more chaotic when reports of a Russian tank breakthrough near Fehrbellin reached the headquarters. Wetzling and the other officers convinced Von Gaudecker that the only solution was to move the division as quickly as possible to the river crossing point. As Russian aircraft circled the area, Von Gaudecker quickly signed an order to the regimental and battalion commanders instructing all units of the division to assemble in the area of Lenzen. The units were to stay intact, and anything pertaining to the V-2 rocket and its components, as well as any converted special vehicles, should be brought along.

That evening as vehicle after vehicle arrived in Lenzen, the battle sounds were all around the division. Fires were visible in nearby villages, and the V-2 division was running out of time. Major Schuetze was in command, as Colonel Von Gaudecker had traveled to the 41 Army Corps headquarters to inform them of his decision to surrender to the Americans; there he was promptly arrested. Schuetze had not arrived in Lenzen because of car problems, and absent of any divisional commander present, none of the regimental commanders wanted to take responsibility to order the columns forward to the Elbe at Wootz. Already security patrols had passed by and asked about the destination and purpose of the movement. A fictitious explanation was given that seemed to justify the move, but it was evident the patrols were becoming very suspicious. As night fell, it was clear no movement could take place before the early morning hours of May 2.

At dawn the next day, word spread among the V-2 troops of Hitler's death, which was announced overnight. Major Matheis took charge and ordered the move toward Wootz. Colonel Wetzling was in the lead vehicle directing the column to the correct crossing location on the river. Along the way, the vehicles ran into difficulties when a road block was encountered on the way to Wootz. A tree was felled across the road, and the crew of an SS assault gun threatened the column if they did not turn back. Colonel Wetzling dressed down the crew and shouted a sharp order for them to get out of the way. It worked. As the assault gun crew moved away, the tree was removed and the col-

umn continued to the river. The crossing from Wootz to Gorleben was made throughout the day without any more complications. Storm boats of the U.S. 121st Engineer
Battalion were used to ferry the soldiers, while the Gorleben Elbe ferry moved the vehicles. After some short interrogations, the V-2 division was escorted to the POW camp
at Herford. If not for the efforts of these few division officers such as Colonel Wetzling,
the whole of the V-2 division might have been slaughtered or taken into captivity by
the Red Army.[32]

During these last few weeks, it appears the remnants of the former SS Werfer
Abteilung 500 were not among the V-2 troops surrendering at the Elbe. By the second
and third week in April, most of the former SS 500 units were disjointed. The
Nebelwerfer batteries expended most of their ammunition in skirmishes along the
route toward Berlin. The troops were engaged in small battles against Russian forces
from the east and American forces from the west. Confusion was everywhere, and the
constant threat of air attacks made any movement extremely difficult. Some of the fractured groups intentionally remained "lost" in an attempt to survive the last days, until
they finally surrendered in small groups at various locations.[33]

Thus, with the capture of the V-2 division, the German A-4/V-2 ballistic missile
campaign came to an end.

Speer's records indicated that by the war's end, 5,797 V-2s had been invoiced by the
Mittelwerk; around 316 more were built elsewhere, thus a total of some 6,112 missiles
were built. Test rounds fired at Peenemünde, Heidelager, and Heidekraut accounted
for close to 751 missiles. Various recorded impact lists show approximately 3,173 missiles launched at Allied targets. Factoring in an eight to ten percent failure rate gives
300 additional operational rockets. Adding in the numbers of rockets captured by the
Allies before they could be launched brings the total number of V-2s to somewhere
over 6,100. Of the approximately 6,100 V-2s built, only about 4,000 were used. At the
end of 1945, British soldiers demolished 17,541 tons of scrap V-weapons parts.

The Mittelwerk factory produced some 4,575 missiles between August 1944 and
March 1945. Approximately 40,000 to 64,000 forced laborers worked in the tunnels
over a 20-month period. It is also projected that of the 64,000 prisoners employed in
and around the Mittelbau complex, 26,500 did not survive. More than 11,000 of these
deaths occurred after April 1945 when the camps were evacuated by the Germans in
the face of the American advance. The evacuation was especially barbaric. The SS
guards shot prisoners, herded them into barns and burned them alive, left them to die
if they were too sick to walk, or made them part of walking or rail convoys headed to
other concentration camps. Over 25,000 of these were killed either by beatings, starvation, and sickness or by the brutal efforts of the SS to relocate them before the
Americans arrived. It is now recognized that many of the most appalling concentration
camp images, which have been embedded into our consciousness from the Second
World War, were taken by U.S. troops as they entered the Mittelbau camps.[34]

Elsewhere, von Braun and the contingent of Peenemünde scientists had moved
from Oberammergau to the resort hotel, Haus Ingeborg, in the border town of

Oberjoch, near Austria. There, von Braun met up with General Dornberger from Peenemünde. Von Braun's brother Magnus was also there. Dornberger told von Braun that General Kammler had been ordered by Hitler to kill all of the Peenemünde scientists and technicians before they were captured by the Americans. Patton's army was still far away, as the supply of fuel (or lack of it) for the Allied columns was slowing the advance of the Americans. Needing food and supplies, the scientists again used the VABV ruse to requisition the items from army supply posts.[35]

There was not much to do except wait for the Americans. The scientists played cards and listen to the radio. They heard of the fall of Berlin on May 1, along with the news that Hitler was dead. As the Americans finally drew near, it was decided that Wernher von Braun's brother, Magnus, would go out to greet the troops and surrender for everyone. The reasoning for this was that Magnus could speak broken English, and it was thought that a large group of German men marching toward the Americans would seem hostile or threatening.

Young Magnus pedaled off on a bicycle to meet the Americans. The first soldier that he encountered was a sentry with the 324th Infantry Regiment, 44th Infantry Division: PFC Frederick Schneikert. Magnus was ordered to drop the bicycle and come forward with his hands up. In a smattering of English laced with bits of German, Magnus tried to explain his mission. The young soldier was not really sure what to do with this boyish figure claiming to be a rocket scientist, so he turned the matter over to his commanding officer, First Lieutenant Charles L. Stewart. Stewart at first thought that Magnus was trying to "sell" his brother and the other scientists to the Americans. The communications were soon cleared up, and Lieutenant Stewart gave Magnus passes for the Germans to ensure their safe passage to the American encampment.[36]

Wernher von Braun, General Dornberger, and several other scientists were so excited after Magnus returned that they piled into three vehicles and immediately headed for the American camp. The Americans were struck by Wernher von Braun's young good looks and his charm. He did not look the part. He did not resemble the imagined image of a top German rocket scientist. The Americans soon realized the importance of their prize. Reporters and newspapers flooded in to see the rocket scientists.

When SS General Kammler dropped out of sight on April 17, 1945, it seems that no one was looking for him. The Allies didn't seem concerned, maybe because they did not realize the full scope of the powers given to Kammler in the closing days of the war. In an attempt to absolutely undermine Speer's authority, Heinrich Himmler appointed Kammler as Special Plenipotentiary for the Armament Industry, a move which was readily agreed to by Hitler. It was a desperate bid, albeit far too late, to accelerate the production of the Me-262 jet fighter to counter the waves of bombers over Germany. The order, which subordinated Speer and even Göring to Kammler, was signed by Hitler in the bowels of the Führer Bunker under the garden of the Reich's Chancellery in Berlin on March 27, 1945. Several weeks later on April 30, 1945, Hitler committed suicide in the lower part of the bunker. Curiously, when Kammler disappeared, the authorities in Germany were not searching for him either.

In March and April, Kammler was constantly on the move, jumping from the Rhineland to Berlin, then to Thuringia, but his headquarters was still located at Suttrop bei Warstein.

Several differing accounts exist of Kammler's whereabouts at the end of the war. It is impossible to discern positively which version might be correct. Officially, his last known location was Bad Essen. Although details are scanty, two of the four accounts could be partially accurate. The most credible account tells of his emergence at Ebensee in Austria.

Photographed on May 3, 1945, after their surrender to U.S. 44th Infantry Division; Lt. Colonel Herbert Axter, Major General Walter Dornberger, and Dr. Wernher von Braun pose for the camera. (*U.S. Army Signal Corps*)

His presence there is substantiated by a postcard mailed to a relative in Germany during late April of 1945. A few years earlier, Kammler had begun a series of underground tunnels near Traunsee Lake. Ebensee was planned as a secret rocket facility and secondary refuge for Peenemünde personnel, but in 1944, the function changed to an underground oil refinery and storage facility. Still, the SS construction office, headed by Kammler, maintained a sub office at Ebensee up until the last days of the war. If Kammler was in Austria during late April, it seems that he might have traveled to Prague and then to the Skoda armaments factory at Pilsen in early May. The SS research offices for advanced weaponry were located at Pilsen and were under the direction of Kammler. Dr. Wilhelm Voss, who was the Skoda complex director, reported that following the U.S. Third Army capture of the factory on May 6, he was told Kammler had committed suicide in a forest between Prague and Pilsen on May 9, 1945. No corpse or grave has ever been found.[37]

By all accounts, Kammler should have been the most wanted man in all of Germany. He was the man responsible for the construction and operation of the Reich's concentration camps and slave labor force, personally responsible for the deaths of thousands, as well being a senior member of the Gestapo leadership. These reasons alone were more than enough to make him a coveted prize to the Allies, but his position as leader of Germany's secret weapons programs, where he was in charge of all predecessors of modern-day armaments, should have made his capture or exhumation imperative. If by chance Kammler managed to somehow survive the war, his hypothetical appearance in late 1945 would have probably been very awkward for Allied officials. He certainly could have been charged with war crimes for the slaughter of Jews and slave laborers, but it remains to be seen how Nuremburg officials could have indicted him for the bombardment of London and Antwerp with terror weapons, especially since the Russians, British, and Americans were already exploiting the captured German technology in their own arsenals.

V-2

The end of WWII brought with it new endeavors and new technology. The V-1 flying bomb, itself an elementary cruise missile, and the V-2 ballistic missile, the forerunner of intercontinental ballistic missiles, confirmed the possibilities of long-range rockets for warfare and space travel. The frenzied rush for Nazi technology and scientists in the closing days of the war created an air of international competition that foreshadowed the coming Cold War. The western Allies, particularly the United States, raced to acquire as much German military technology as possible before the Russians could take possession of certain sectors of Germany, which had been designated to them by the Yalta agreement. But as it turned out, the Soviets acquired their fair share of prominent German scientists and technology.

When the Americans questioned the German scientists about their advancements in rocketry and propulsion, they stared back at their interrogators in bewilderment. "We were only expanding on the work of the American scientist Robert Goddard, why haven't you asked him about rocketry?"

9

Legacy

In America, Dr. Robert Goddard had spent the war years doing experimental work under contract for the United States Navy. Earlier, Goddard had become alarmed when the periodical inquiries he had received from German scientists abruptly fell silent in 1939. It was the country of Nazi Germany that had shown the greatest interest in Goddard's work prior to the outbreak of world war. Could the Germans be developing a rocket for use against Allied armies or cities? Had his gullible communications with German scientists, along with the patents he had filed, given them enough information to capitalize on his work? Concerned about Hitler's possible use of rocket weapons, and before the United States formally entered the war, Goddard traveled to Washington to meet with Army officials to discuss the possible use of liquid-fueled rockets as a new form of long-range artillery. After showing some films, which highlighted a few of his recent rocket flights from New Mexico, Goddard was politely thanked and then shown the door. The U.S. Army was not interested.

At the end of the Second World War, Goddard returned to his private experiments with rockets. His first goal was to inspect the huge German V-2, which he had heard so much about. When he finally was allowed to examine the inner workings of the German missile, only weeks before his death from cancer, Goddard commented, "It looks like one of my rockets." Although the combustion principles of the V-2 were inherent in one of Goddard's designs, the German missile was far more advanced than even Goddard imagined.

The race to discover and capture the secrets of the German missile began even before the hostilities in Europe ended. On April 10, 1945, all work had stopped at the Mittelwerk underground factory. More than 4,000 personnel, including plant officials, civilian specialists and engineers, and the military guard had fled the factory. The next day, the spearhead of the advancing American troops, Combat Command B (CCB) of the U.S. Third Armored Division, under Brigadier General Truman Boudinot, entered Nordhausen. Here CCB was to pause and link up with the U.S. 104th Infantry

"Timberwolf" Division before continuing its drive to the east. One of the first sickening encounters with genocide took place at Boelcke-Kaserne, a former German military barracks that was used by the SS as a dumping ground for prisoners from Mittelbau sub camps who were too weak or diseased to include in the transports or forced marches out of the area. An estimated 1,300 to 2,500 corpses were found there lying in the sun. The dead also included prisoners killed in an Allied bombing raid aimed at Nordhausen.[1]

Several miles further, as they approach the foothills of the Harz Mountains, they met several skeletal figures wearing ragged, striped uniforms who told of "something extraordinary" inside of the mountain. Moving further, the American troops discovered Dora and the entrances to the Mittelwerk tunnels. When walking into the first long tunnel, they were stunned to see railway freight cars loaded with the mammoth V-2s. As they continued further into the depths of the mountain, they found the entire workings of a gigantic factory, complete with precision heavy machinery and missile parts, all laid in neat rows of assembly. The assembly line had been abandoned with electric power and ventilation systems still running. Walking through the illuminated tunnels, Major William Castille, Intelligence Officer for CCB, was quoted as saying, "It was like being in a magician's cave."

When word came of the incredible find, U.S. Colonel Holgar Toftoy, Chief of Army Ordnance Technical Intelligence, immediately began arranging for "Special Mission V-2" from his office in Paris. Its purpose would be the evacuation of 100 complete V-2s and specialized parts back to the United States. To support his mission, Toftoy had organized special rapid-response Ordnance Technical Intelligence teams attached to each Army group. These teams were equipped with cameras, radios, transport trucks, and qualified personnel whose job it was to ferret out interesting weapons technology and record it. The team designated to investigate the Mittelwerk was headed by Major James Hamill of Ordnance Technical Intelligence. He was assisted by Major William Bromley in charge of technical operations and by Dr. Louis Woodruff, an MIT electrical engineering professor, as special advisor. The team was headquartered in Fulda, about 80 miles southwest of Nordhausen.

After rounding up captured German rolling stock and clearing a path into the tunnels, Special Mission V-2 succeeded in loading up and sending off its first 40-car trainload of V-2 parts and machine equipment on May 22, 1945. Nine days later, the last of the 341 rail cars left the Mittelwerk bound for Antwerp. Although the British properly protested that by prior agreement, half the captured V-2s were to be turned over to them, the Americans ignored these protests. Sixteen Liberty ships, bearing the components for 100 V-2 rockets, finally sailed from Antwerp, destined for New Orleans and then White Sands. Hamill was not told that the factory would be in the Soviet zone of occupation. Consequently, quite a number of missile parts were left for the Soviets to discover.[2]

Major Robert Staver from the Rocket Section of the Research and Development branch of the Ordnance Office was tasked in directing the effort to find and interro-

Southwest exit from Tunnel B (left) at Mittelwerk near Nordhausen. Allied transports moved hundreds of tons of mate-rial from the facility before handing it over to the Russians. Special Mission V-2 succeeded in loading up and send-ing off its first 40-car trainload of V-2 parts on May 22, 1945. V-2 rocket (right) in the tunnels near Nordhausen. In this photo it is apparent just how much of the fuselage was occupied by the large fuel tanks. This particular rocket was partially disassembled for inspection soon after the underground Mittelwerk manufacturing plant was captured in 1945. (*NARA*)

gate the German rocket specialists who had built the V-2. Since April 30 he had been in the Nordhausen area searching the smaller laboratories for V-2 technicians. On May 12, Staver located his first V-2 engineer, Karl Otto Fleisher, who began to put him in touch with other Mittelwerk engineers who had not been part of von Braun's caravan to Bavaria. On May 14, Staver found Walther Riedel, head of the Peenemünde rocket motor and structural design section, who urged the Americans to import perhaps 40 of the top V-2 engineers to America.[3]

After the general German surrender on May 7, 1945, the V-2 specialists and engi-neers were moved to a prisoner enclosure in Garmisch-Partenkirchen, where a variety of Allied interrogators questioned them. At this point the Americans had the missiles, they had the scientists, but they were still missing the all-important Peenemünde doc-umentation. Fourteen tons of Peenemünde documents had been hidden in an aban-doned iron mine in the isolated village of Dornten in early April. Von Braun had ordered the documents hidden to prevent their destruction by Kammler, and to also use them as a bargaining chip in negotiating their fate with the Allies. As it happened, Karl Otto Fleisher was the only person remaining in the Nordhausen area who was aware of the general location of the V-2 documents hidden by von Braun's group. Staver succeeded in tricking him into believing that von Braun had already authorized him to reveal the location of the papers, and on May 20 Fleisher revealed the Dornten tunnel. In less than a week, the Dornten area was scheduled to fall into the hands of the British. A frantic scramble then ensued to transport the documents back to Nordhausen, where they were quickly shipped to Paris, and then to the Aberdeen Proving Ground in Maryland.[4]

At the beginning of June, Staver requested that some of von Braun's senior engi-neers be sent back to the area of Nordhausen to help identify which of the thousands of German technicians should be offered evacuation to the American zone before the scheduled handover of Nordhausen to the Soviets on June 21, 1945. On June 20, some 1,000 German V-2 personnel and their families were gathered up and placed aboard a long train, which finally made it to the small town of Witzenhausen, some 40 miles to the southwest and just inside the American zone.[5]

This was the beginning of Operation Overcast, later renamed Operation Paperclip. After realizing the tremendous strides made by the German scientists in the field of guided weapons, American military officials wanted to seize that knowledge and incorporate it into their future arsenal. It was imperative to retain the data acquired by the Germans and to go forward. Approximately 150 of the top technicians were rounded up, and after preliminary interrogation and background investigations, they were offered five-year contracts to come to the United States and work for the U.S. Army. In return, the Americans promised to provide housing for their families, who would remain in Germany until provision could be made to bring them to the United States.

At the close of the war in Europe, the Americans were not the only ones hunting for rocket booty. After the German surrender, the British attachment of the Allied Air Defense Division became known as the Special Projectile Operations Group (SPOG). The Air Defense Division had been charged with the task of collecting information about the secret German rockets to formulate effective countermeasures before the close of the war. By April of 1945, the German rocket troops were retreating and London was licking its wounds. Although much was known about the V-2, most of this information had been gathered through intelligence. When the wayward V-2 crashed in Sweden late in 1944, the British government gained a small glimpse of the German missile after securing the wreckage. From this mangled mass, British intelligence obtained a small amount of information, but the Allies could only speculate how the missile was transported, fueled, and fired during hostilities.

After the hostilities ended, more than 8,000 German rocket troops had been captured along with hundreds of Peenemünde scientists. A proposal was put forward by J. C. C. Bernard, A.T.S., Personal Assistant to Major-General Cameron, head of the British Air Defense Division, that the German rocket troops be forced to demonstrate their V-2 handling and firing procedures by actually preparing and launching some V-2 rockets. In June General Eisenhower sanctioned the series of tests, and the Air Defense Division was given the go-ahead with the procedure under the new organization of Special Projectile Operations Group. The project would be known as "Operation Backfire," a comprehensive investigation conducted by the military to completely evaluate the V-2's innovative technology, to interrogate German personnel specialized in all phases of its operation, and then to actually launch several missiles across the North Sea. The program was under the command of Major-General Cameron, along with SPOG General Staff Colonel W. S. J. Carter, who was in charge of operations and documentation.[6]

The gun-testing range at Altenwalde near Cuxhaven, Germany, which was in the British zone of occupation, was chosen as the appropriate testing site. The British utilized the hangars and other facilities of the former German Navy artillery range for handling, logistics, and preparation of the missiles. This location was found suitable because of the sea to the north, with good radar tracking points downrange. A large concrete firing point was created at the site so that there was ample room for the operations, which were to be photographed and filmed extensively.

The captured German rocket troops were fairly willing to demonstrate their V-2 firing procedures, and soon 200 Peenemünde scientists, 200 V-2 firing troops, and 600 ordinary POWs were transported to Cuxhaven. Upon arrival, they were split into two groups and interrogated. The information given by each group was then compared. Wernher von Braun and General Walter Dornberger were also brought to Cuxhaven, but they were not taken to the actual firing site. A comprehensive evaluation of the V-2 had never been undertaken, not even by the Germans. German security would not allow such broad coverage, feeling that no single person should know more about the entire system than the absolute minimum required for their own duties.

Allied air attacks and German sabotage, not to mention the American heist at the Mittelwerk, had consumed almost every assembled V-2 in Europe. Therefore, finding intact V-2s was a great problem. A small British advance party, led by Colonel T. R. B. Sanders, secured several missiles outside of Nordhausen before the Americans arrived in the area, but more were needed. The Americans later removed enough parts from the underground Mittelwerk facility to assemble 70 to 100 rockets in the United States. After the Americans were finished, the British were given another opportunity to salvage what was left before the Russians took over the Mittelwerk. In addition, a small group of intact V-2s was later discovered on an abandoned railway shipment in the British zone of occupation. Officials determined they had enough rockets and parts to assemble about eight rockets for testing. However, they also found that they were missing several key rocket components as well as the support vehicles needed to fuel and launch the rockets.

What followed was an amazing search all over Europe for the missing items. Search parties were sent out everywhere with soldiers who were fluent in German, each with a convoy of trucks, to seek out the missing parts. When the hunt was finished, 400 railway cars and 70 Lancaster flights were used to bring the 250,000 parts and 60 specialized vehicles to Cuxhaven; the most elusive components were the batteries that powered the guidance gyros. Intact tail units were also difficult to locate, and following a request from the British, some of the tails were shipped back from the United States. The explosive material in the missile warheads was steamed out and replaced with sand. At Altenwalde, a giant tower was constructed out of Bailey bridge panels for testing of the rockets in the vertical position. More than 2,500 additional British troops were brought in to complete the various construction projects.[7]

By the beginning of October, 1945, the British were ready to begin. The rebuilt rockets were painted in a black-and-white checkered pattern, similar to the early Peenemünde rocket schemes. On October 1, the first V-2 launch attempt failed due to a faulty igniter. Another attempt was readied, and on October 2, 1945, the first successful test launch of the V-2 by the British soared across the Baltic, almost three years to the day of the first successful German test at Peenemünde. The second successful launch occurred on October 4, followed by the third and final launch on October 15, 1945, with officials from several nations present. The V-2 performed flawlessly and landed near its target point in the North Sea. The operation was concluded a few weeks later.

Some feel this whole operation was as much about convincing the German rocket scientists to come to Great Britain and work for the British in the development of a rocket program as it was about testing the V-2 systems. The British and Americans began fighting over the German scientists even before the war's end. The Americans agreed to "lend" many of the top German rocket personnel for the Backfire tests and later found that the British were trying to convince the Germans to stay after the tests. It took a considerable amount of prodding by the U.S. War Department to gain the return of many Germans to American custody.[8]

Eventually, the British agreed to return the Germans on the condition that four or five of the top scientists could be fully interrogated in London about technical information. As it turned out, there was never a technical discussion for the Germans in London. Von Braun, Dornberger, and several other Peenemünde department chiefs were driven through the streets of London to show them the destruction that the missile had wrought. Von Braun and the other chiefs were returned to American custody in Germany soon after, but General Dornberger was kept by the British. Dornberger probably knew more about the V-2 organization and systems than anyone else, but the British didn't want his technical knowledge, they wanted to execute him. Since Kammler was nowhere to be found, the British fully intended to bring Dornberger to trial at Nuremberg as the person responsible for the bombardment of London with V-2s. It was not until 1947 that Dornberger was quietly released from British custody. It would have been hard to make a case against him in light of all the deaths caused by Allied bombing raids on Japanese and German civilians during the war.

Watching each of the three V-2s that rose from a launch site at Cuxhaven was a certain Russian Army colonel named Sergei Korolev. He had been a part of the Soviet intelligence team sent to investigate battered Peenemünde following its capture by Russian forces and was responsible for the exploitation of what remained at Nordhausen. Only ten years later, Korolev would be championed as the Soviet Union's chief spacecraft designer and the grandee responsible for building the Vostok, Voshkod, and Soyuz spacecraft, which since the 1960s have carried all Soviet cosmonauts into orbit.[9]

Meanwhile, von Braun and his team were heavily interrogated and jealously protected from Russian agents. Dornberger later told British interrogators, "The Russians sent one of my former engineers to me when I was with the Americans, who told me he had an offer to make on behalf of the Russians. We were to go back to Peenemünde and it would be rebuilt, along with a parallel factory in Russia, and they offered to pay us double what the Americans were offering us. We also could move our families with

Captured German rocket equipment at White Sands Proving Ground in New Mexico. The V-2 is on a Meillerwagen trailer next to the 15-ton-capacity Fries Strabo Crane. In June of 1945, the United States government approved the transfer of German rocket engineers to America under the auspices of what became known as Operation Paperclip. Wernher von Braun and more than 100 Peenemünde specialists were transferred to Fort Bliss, Texas, a large Army installation just north of El Paso. Here they trained military, industrial, and university personnel in the intricacies of rockets and guided missiles and helped refurbish, assemble, and launch a number of V-2s that had been shipped from Germany to the White Sands. (Author)

us, etc. We turned it down flat. Afterward, in the town Witzenhausen, they tried to kidnap our leading men such as Dr. Wernher von Braun. They appeared at night as British soldiers in uniforms; I guess they didn't realize it was the American zone. Somehow they had obtained a proper pass, but the Americans quickly realized what was happening and sent them away. That's how those Russians operated, real kidnapping, they had no scruples at all."[10]

A few of the German scientists were disappointed with the American offer of employment in the United States. When the Soviets began broadcasting offers to German engineers using large megaphones, several of the Germans accepted and crossed over to the Soviet zone, most notably Helmut Gröttrup. Gröttrup, who claimed he was upset with the American offer because his family would have to remain in Germany, had never really gotten along with Wernher von Braun. Gröttrup was eventually named head of the new rocket institute at Nordhausen, which was established by the Russians the following year. The Soviets allowed Gröttrup to build up his research institute until October 22, 1946, after which they forced the Germans to move east to Russia along with thousands of other specialists from Eastern Europe. The most experienced German specialists were ordered to board trains and were sent to various locations throughout the USSR to assist in the organization of missile production and design. By the beginning of the 1947, the Soviets had completed the transfer of all rocket technology from Germany to various secret locations in the USSR. A year later, beginning on October 30, 1947, the Soviet-German team launched 11 V-2 (R-1) rockets near the village of Kapustin Yar, north of the Caspian Sea. The R-1 was manufactured from scratch, as Stalin had ordered that no German-manufactured parts would be used on the Soviet rockets. Once the Soviets had acquired the German knowledge, Stalin's military tossed Gröttrup and his assistants aside in favor of the many capable Soviet engineers. Never fully integrated into the Russian missile program, Gröttrup was eventually sent back to Germany.[11]

Only a small percentage of V-2 personnel from the wartime operational batteries were invited to the Backfire trials. Following their surrender to the Americans at the Elbe, most spent their time in Allied POW camps before returning to their families in Germany in 1946–1947. Similarly, very few members of von Braun's team participated in the experimental launches at Cuxhaven.

The first of the Operation Paperclip scientists arrived at Fort Strong, New York, on September 20, 1945, and then were moved to Aberdeen Proving Ground in Maryland. The study at Aberdeen was centered on the processing of captured guided-missile documents. Here, the Americans and Germans went over thousands of documents, putting them in order and creating accurate translations of the most important. By this time, most of the other Peenemünders had moved in at Fort Bliss, near El Paso, Texas. The components for the V-2s had been shipped from Germany to the desert near Las Cruces, New Mexico, and then to the newly established White Sands Proving Ground, where the work of assembling and launching test missiles for the United States government began. By February of 1946, almost all of von Braun's Peenemünde team had been reunited at White Sands; and on April 16, the first V-2 was launched from New Mexico. The U.S. ballistic missile program was underway.

From 1946 to 1952, the United States launched more than 60 V-2s. The once-deadly V-2 was turned into a scientific investigation platform, with its nose and control compartments stuffed with high-altitude research projects. A variant of the V-2 would be known as the "Bumper." This consisted of the first-stage V-2 mated with the smaller second-stage Army WAC Corporal on top. This two-stage missile configuration would reach altitudes never before attained. The need for more space to test fire bigger missiles quickly became evident. In 1949, the Joint Long Range Proving Ground was established at what was then a very hostile and remote area in Florida named Cape Canaveral. On July 24, 1950, the Bumper configuration became the first of hundreds of test missiles to be fired from Cape Canaveral. Around the same time, the Army's missile program was transferred from White Sands to a location just outside of the small town of Huntsville, Alabama. Here, von Braun's team worked to develop the Redstone rocket, named for the U.S. Army arsenal where it was born. The German scientists would call Huntsville home for the next 20 years.

The Army Ballistic Missile Agency was established at Redstone Arsenal to craft the Jupiter intermediate range ballistic missile. The Jupiter-C, an updated version of von Braun's Redstone rocket, was launched on January 31, 1958, carrying with it America's

The Bumper project was designed to study the possibility of a two-stage research missile consisting of a V-2 first stage and a WAC Corporal as the second stage. It would be the first two-stage liquid-fueled rocket ever built, and as such would explore problems related to stage separation and rocket ignition at high altitude. The V-2 flew an arched trajectory and released the Bumper WAC in a horizontal attitude. Bumper V-2 was the first missile launched at Cape Canaveral on July 24, 1950, and eventually reached a speed of 3,270 miles per hour. (*NASA*)

first satellite into orbit, the Explorer I. Three years later, the Mercury-Redstone launched Alan Shepard and Virgil "Gus" Grissom on suborbital space flights, paving the way for John Glenn's first orbital flight.

Following the establishment of The National Aeronautics and Space Administration (NASA) in 1958, the entire Army Ballistic Missile Agency was transferred to NASA to become the nucleus of the new agency's space program. The Marshall Space Flight Center pursued a separate course from the Army's development of military rockets. The newly named research center concentrated on the development of U.S. rockets for space exploration. The series of huge Saturn rockets were developed at Marshall to support the Apollo program and to honor President Kennedy's pledge of taking man to the moon. They were the most powerful space launch vehicles the world has ever seen. Former Peenemünders were on hand in Mission Control when, on July 20, 1969, a transmission from the moon's Sea of Tranquility reported, "The Eagle has landed."

After leaving NASA, von Braun was active in establishing and promoting the National Space Institute; but not long after, doctors revealed he had cancer. He died on June 16, 1977, in Alexandria, Virginia.

Many today believe that von Braun should not have been celebrated as a hero. They feel he turned a blind eye toward what was happening to slave laborers in Germany. There is not much evidence in official records to indicate that he was even somewhat troubled by the use of slave labor. The means through which his experimentation could continue was the vehicle of the Wehrmacht's war machine. While all of this may be true, von Braun's work on the rocket was his passion throughout his whole life, not just during the war. His crime was one of success through available means or the greater evil of complacency, which infected so much of the German population during WWII.

Following his release from British custody in the summer of 1947, Walter Dornberger traveled to the U.S. zone in Germany, where he worked a stint for the American authorities. In 1950 he was hired as a consultant for the Bell Aircraft Company and later became the driving force behind NASA's Dyna-Soar project. The project borrowed much from Eugen Sänger's Silbervogel bomber, which was another Nazi wartime endeavor. Were it not for its cancellation in 1963, the Dyna-Soar, otherwise known as the X-20, would have reached hypersonic speeds and returned through the atmosphere to land and be used again, making it a forerunner of the modern space shuttle. Dornberger rose to the position of Senior Vice President at Bell Aerosystems before he returned to Germany following his retirement in 1965. He died in 1980 in Baden-Württemberg.[12]

Since the end of WWII, the historical account of the German V-weapon campaign has been lambasted in many pieces of literature as a total waste of time, effort, and resources. It is clear that the V-2, just like any other solitary German weapon, could not have turned the tide of war in Germany's favor in 1944–1945. The mighty arsenal of democracy rolled over a multitude of German weaponry, but the V-2 is most often singled out as the biggest failure of the entire Nazi arsenal, simply because it failed to win the war for Germany; quite a lofty standard for any weapon of the period. This determination is usually accompanied with the contradictory exaltation of the ballistic missile's potential in future conflicts.

In reality, the deployment of the V-2 rocket greatly affected the Allies' strategic planning. When the information about the V-weapons first came to light, the threat alone was enough to cause panic. During Operation Crossbow, the Allies were forced to divert thousands of aircraft in what turned out to be an ineffective attempt to stop the V-weapons. These aircraft could have been dropping their bombs, hundreds of thousands of tons, on critical German infrastructure and military targets in the heart of the Third Reich. Allied leaders feared that the V-weapons could disrupt the D-Day invasion or the reinforcement of the beachheads at Normandy. It was imperative that Crossbow delay the threat long enough to capture the beaches and secure a foothold. Estimates about the destruction and death toll caused by the combined V-1 and V-2 attacks are horrendous. During a nine-month period, the V-weapons destroyed a combined total of 37,000 homes in England and Belgium and damaged approximately 1.5 million others. The missiles killed almost 9,000 people and wounded 25,000. The attacks also affected British industry, as a significant portion of English armament production was moved from the city of London to outlying areas to reduce the risk of damage from falling V-bombs. When the V-2s began falling on London in September of 1945, Field Marshal Montgomery, in an attempt to cut off the launching sites on the Dutch coast, altered his plan of attack, which was originally slated to cross the Rhine River near Wesel. The failure of Operation Market Garden at Arnhem can be partially attributed to the would-be missile threat. In addition, the bombardment of the Antwerp harbor slowed the off-loading of supplies significantly, resulting in decisions that ultimately prevented the port from ever becoming completely operational. There was not much the Allies could do to stop the rocket attacks. The V-2s were effective simply because they traveled at such great heights and at great speeds that shooting them down was impossible. The only successful counter to the V-2 was the eventual overrunning of the missile launching sites, which finally pushed the firing crews out of range.[13]

Many historians pan the deployment of the V-2, quoting the statistics of the missile's inaccuracy, saying it could only be used against a large target such as London, but in their next breath assert how the V-weapons "paved the way for modern missiles and tactics." Obviously, if the V-2 had been able to strike a target with pinpoint accuracy, it would have been considered a triumphant success. However, the Western Allies possessed no such pinpoint weaponry.

The German military leadership lacked a specific operational strategy for the employment of the V-weapons. After the V-2 surpassed most of its technical and bureaucratic obstacles, Hitler's only aspiration was to launch hundreds of missiles each day at London. His motive was solely retaliation, not strategic. The continuous Allied bombardment of German population centers made him even more determined to take revenge. He wanted new offensive weapons that would bring the war to British civilians. He continued to give the necessary resources for the successful manufacture of vengeance weapons, and it was the threat of "what" the V-2's warhead might have contained that commanded so much attention from the Allies. Hitler opted not to employ deadly gas or nerve agents in the missile's warhead for fear that Allied bombers would reciprocate over German cities. What if 50 missiles, each carrying warheads filled with poison gas, chemical weapons, or even a dirty bomb of radiological material had landed in the embarkation areas of southern England in the weeks leading up to D-Day? It is easy to imagine the havoc that could have been wrought by the unstoppable V-2s in such a scenario. Even General Eisenhower hinted that if the V-2 had been operational several years earlier and tens of thousands of missiles, armed with only conventional warheads, had rained down on English ports, the Allied assault of Fortress Europe might have been prevented.[14]

Since 1945, the V-2's reputation for inaccuracy has been continually perpetuated. While this is certainly true from a tactical perspective, the causes are not usually fully explored. The system was vulnerable to many factors causing deviations to the planned flight path. Was the wide dispersion only due to inadequate system design or system failures? Probably not. Like any artillery weapon, the Germans needed to know the exact range and location of the target. In 1944, the accuracy of the European geodetic surveys left much to be desired. Even the exceptional national survey grids of Great Britain were so poorly linked with continental grids that this alone accounted for error in the missiles of approximately 1,000 to 2,000 feet. It is very possible that coordinate transformation, if either the distance or the direction on the map is wrong, along with the correction for the earth's curvature and rotation, played a greater role in the rocket's dispersion than did the limited guidance system.

Also, German reconnaissance flights over England were practically nonexistent. The lack of coexisting intelligence forced the Germans to depend on their agents within England for information. British Intelligence had locked up most all of the German spies in London. Some of the agents were even turned around by the British to feed back bogus information about the impact locations. At best, the Germans could only disregard information from their informants—at worst, they believed them. The Germans resorted to reading the obituary sections of the London newspapers to establish the location of impacts. Soon the amount of information reported in the papers was restricted by British authorities. Accurate spotters could have greatly improved the adjustments made for targeting, but the German rocket crews never received this feedback in reality.[15]

If the V-2 is to be deemed ineffective for its lack of accuracy, one must examine in comparison the accuracy of other forms of aerial bombardment. Allied strategic bombing was certainly not a precise means of delivering munitions, and it too could only be effective if used against large cities and infrastructure. Even operating in the best of conditions, during WWII, Allied heavy bombers could place only 70 out of 500 bombs within 1,000 feet of an intended target. This meant that a flight of 100 B-17s, each dropping five 500-pound bombs, and risking the lives of 1,000 airmen, could only expect a few bombs to actually hit the target. More than 9,000 bombs would need to be dropped to have a 90 percent chance of hitting a target the size of a small factory. Operating some five miles high, the daylight bombers of the Eighth Air Force sometimes hit the wrong cities or even the wrong countries. They often bombed open country and, more often than not, missed their intended targets by hundreds and thousands of yards. America's own clandestine wonder weapon, the acclaimed Norden bombsight, was practically useless over cloud-covered targets. Moreover, as Luftwaffe fighters and radar-directed antiaircraft defenses took an increasing toll on American bombers, the aircraft were flown even higher to trim down losses, which further degraded the bombing accuracy. It is no wonder that German war production continued to grow until very late in the war.[16]

Some have argued that the V-2 was not an economical method of delivering bombs on the enemy, citing that a single Allied bombing raid of WWII had the ability to drop more tons of explosives than the V-2 delivered during its whole operational period. When comparing total pounds of explosives delivered by the V-2 as opposed to the Allied bombing campaign, historians never take into account the supersonic impact velocity of the missile. Compared to the ordinary bomb, the effects of the V-2 impacts were extraordinary. Not only did the rocket consist of a warhead weighing almost

2,000 pounds, but it also impacted the earth at 3,000 feet per second. The inertia of several tons of metal striking the earth at this speed could cause significant damage even without the following explosion of the warhead.

It is certainly possible that the German military discovered the truth about strategic bombing during the Battle of Britain. The cost versus the effectiveness of building and maintaining a manned heavy bomber fleet, in the face of profound Allied air superiority, would have definitely been more expensive than the possibilities of a ballistic missile program.

Dornberger wrote in his memoirs, "Since the enormous loss of bomber planes as a result of the attacks against England in 1940, my colleagues and I have been firmly convinced that a defeat in the air war on the western front could be prevented, if at all, only by the employment of guided missiles of very great range and effect."

The V-bombs were in fact conceived as replacements for the fable of strategic bombing; anyone who doubts this need only turn on his or her television today.

Maybe Dornberger and von Braun did not develop the V-2 too late in the war, on the contrary, possibly it was developed too early to live up to its expectations. Even if the missile had been deployed a few years earlier, it is doubtful that it would have changed the eventual outcome of the conflict. Still, this does not mean the V-2's deployment was ineffective. On the other hand, it was not until a few decades later, with the arrival of advanced guidance systems controlled by computers and missiles topped with atomic warheads, that the ballistic missile concept gained true credibility.

Critics of the V-2 often list the alternative uses for the manpower and material allotted to its development and production. The diversion of Germany's best engineers and technicians to solve the missile's complicated technical troubles is frequently linked to the lethargic development other critical weapons systems. Repeatedly, the comparison of the total construction of over 30,000 V-1s and 6,000 V-2s is equated to the production of 24,000 more Luftwaffe fighters. Those who hold this point of view fail to appreciate Germany's strategic position late in the war. With Allied aircraft ruling the skies over Europe, the production of additional fighter aircraft, which would have been purely a defensive move, would have made no impact on the air war. In 1942 Hitler was actually scaling down the Luftwaffe on the assumption that the war was more or less won. Little effort was directed toward the development of new aircraft designs. Germany had not developed any offensive heavy four-engine bombers and could only deploy their fighter arm in defense of the Fatherland. But as the air war continued, the available supply of experienced pilots and serviceable aircraft waned. Later, when Speer was able to speed up fighter production, the new aircraft sat idle, either because of the lack of trained pilots or the lack of aviation fuel. The fuels for the V-2— alcohol and liquid oxygen—were still available. In the long run, not only was the rocket Germany's only means by which it could strike back at the enemy, it was far less expensive and far more effective than the production of useless fighter aircraft.[17]

It is certainly clear that Germany's development of the guided missiles panned out well for the United States and the Soviet Union during the later half of the twentieth century. The German rocket development led to enormous changes in the world after the Second World War. The Saturn rockets, one of which took the first men to the moon in 1969, were the direct descendants of the V-2 and were engineered by Wernher von Braun and many members of the same Peenemünde team. V-2 technology also led to the nuclear-armed intercontinental ballistic missiles (ICBMs) developed by the United States and Russia in the late 1950s.

Sixty years after the Allies' struggle to counter the German missile threat, the lessons they learned still apply today. The fear of warheads containing poison gas, chemical agents, and even nuclear weapons was evident during the Gulf War of 1991. Iraqi president Saddam Hussein ordered the launch of SCUD missiles against Israel in a purposeful attempt to draw the Israelis into the conflict, in the hopes of prompting Arab forces to withdraw from the coalition. The SCUDs struck in the same indiscriminate way as Hitler's vengeance weapons. The operational firing procedure for the SCUD missile was almost identical to that of its German ancestor. The MAZ-543 Transporter-Erector-Launcher moved from its hiding place and traveled to a rendezvous site where the rocket was mounted. The fueling trucks drove to the launching point, and the rocket was oriented, fueled, and fired. Iraq attacked with approximately 88 SCUDs, almost all of them Al Hussein models, with 46 striking in Kuwait and 42 in or near Israel. Like the V-2, several SCUDs failed and crashed close to where they were launched and a substantial proportion of the Al Hussein SCUD models spontaneously broke up during reentry. The Iraqi warheads contained only conventional explosives, and in their wisdom, Israeli leaders chose to withhold retaliation to secure the stability of the United Nation's coalition. However, the coalition air forces expended a significant amount of energy attacking related SCUD targets; energy that could have been better directed possibly somewhere else in the theater. Like their predecessors of 50 years earlier, the Allied leaders of Operation Desert Storm determined the best way to counter the missile threat was to destroy the mobile launchers on the ground. In both cases, Allied aircraft owned the skies over the enemy's territory; nonetheless, the missiles were no easier to find in the Iraqi desert than they had been in The Hague.[18]

Today, any target that can be seen from the air, or precisely located with the Global Positioning System (GPS), can be struck with a single precision bomb. Four B-2 stealth bombers, each carrying 16 2,000-pound bombs can destroy several dozen strategic targets in one raid. It is estimated that on the first day of Operation Iraqi Freedom, more than 700 Coalition aircraft flew missions against more than 100 targets in Iraq, dropping precision-guided munitions on numerous military targets. The Coalition dropped over 6,000 bombs and launched 600 Tomahawk cruise missiles in the first ten days of the war. The V-1 flying bomb was proven to be the predecessor of the modern cruise missile. Today's cruise missiles, such as the Tomahawk, Exocet, and the Russian Mach-2 SS-N-22 Moskit, incorporate GPS for use in the guidance controls and can carry a warhead tipped with a nuclear explosive. Moreover, their ability to adjust altitude, speed, and course further enhance their effectiveness. The V-weapons survive in the modern arsenals of today. Their range, accuracy, and destructive power have improved significantly; however, their overall offensive capacity has been challenged by vast improvements in defensive technology. These weapons have changed our view of the world and of war forever.[19]

The legacy of the German V-2 ballistic missile represents the best and worst of humankind's endeavors in the twentieth century. As fascinating as it was horrific, historians and scholars have been debating the real consequences of Germany's rocket weapon development program for the better part of 60 years. Psychologically, one might ask how in the negative atmosphere of Nazi Germany was so much achieved on a technical scale in such a short period? Deeper thoughts about the realities of war and man's inhumanity to man can bring a shiver. This is nothing new of course; humans have used invention, cleverness, and cruelty in war since the beginning of time. However, if we conclude that most of our significant technological advancements have occurred during wartime, or as a result of research into weapons of war, then does this imply that we are not the admirable souls we think we are? Maybe.

Appendix

Description of V-2 Maintenance, Supply, and Firing Operations

A typical trailer-mounted rocket was supported by about 30 vehicles, which included a transport trailer, mobile crane, launch table and trailer, propellant vehicles, and command and control trucks. The rockets and warheads, along with the liquid oxygen and alcohol, would be delivered to the firing areas by railway. The deployment of the missiles to the mobile firing sites was certainly not an easy task, but it was a task that the Germans performed quite efficiently. The multitude of troops required seemed to be the single biggest drawback.

The rockets, normally 20 per shipment, were transported to the general firing areas by railway, where they were met by a supply detachment of the technical troop. This was made up of about 100 cars, 62 tractors, 33 dispatch riders, some 22 liquid oxygen bowser trailers, two 15-ton capacity Fries Strabo mobile gantry cranes, 48 Opel Blitz alcohol tankers, four hydrogen peroxide tankers, and four towed pump trailers with sodium permanganate (Z-stoff) off-loaded in barrels. After the missiles had arrived at the railhead, one of the cranes was towed over to the rail tracks by the technical troops and put into position. The train pulled the rockets beneath the crane and then, with the use of a lifting beam attached to two straps around the rocket, hoisted them over to a waiting Vidalwagen, a 46-foot-long lightweight road transporter trailer. The Vidalwagen, a simple A-frame tubular design with custom supports to cradle the unfueled rockets, transported the V-2s to the field stores. The supply detachment personnel would receive the shipments at the railhead and stockpile the materials in the firing area field store.

The field store, normally located a few miles from the actual firing sites, was maintained by the technical crews of the battery. The technical troops would receive the rocket and prepare it for transfer to the firing troops. Tests performed at the field stores included checks of the propulsion systems, steering, alcohol-tank pressurization, and wiring. Any problems that could not be repaired at the field stores meant that the rockets would have to be returned to the Mittelwerk. The warheads were fixed to the rockets at the field store using a small overhead crane to hoist the warheads into place as the technical crews bolted it to the front of the rockets. The "Elefant," a cylindrical drum around the warhead, supported the weight of the warhead during this procedure and protected the warhead during shipment. After the warhead was secure, the transit drum was removed and stockpiled for reuse.

When a V-2 was ready to be launched, the technical troops handed the rocket over to the firing crews. This was done at a rendezvous location somewhere between the

Soldiers of the technical troop (left) attached the warheads to the rockets at the battery's field store, which was located some distance from the firing site. The warheads were shipped inside of large, cylindrical transit drums. Lifted by a small crane, the one-ton transit drum was supported while the warhead was secured to the front of the rocket using a series of bolts. A securing bolt was removed, and the drum was slid off. (*Deutsches Museum Munich*)
The Elefant V-2 warhead transport drum (right). These containers held the 1,700-pound explosive warhead during transport to the launching sites. This example has remained in the forest more than 60 years. (Author)

field store and the firing sites. At the transfer point, at some secluded and prearranged spot in the forest in order to provide as much concealment as possible from prowling Allied fighter aircraft, the mobile gantry crane lifted the four-ton unfueled rocket and moved it from the Vidalwagen road transport trailer onto the frame of the firing crew's Meillerwagen erector trailer. The Meillerwagen consisted of two major components: the transportation frame and the erecting frame. The erecting frame was fitted with two large clamps that gripped the V-2 around the airframe above the fins and also around the warhead. With the rocket secured in the clamps of the Meillerwagen, the Vidalwagen would pull away. The Meillerwagen would be hitched to the towing vehicle—normally the halftrack prime mover Zugkraftwagen 12-ton heavy-traction vehicle or the Wehrmacht heavy tractor Hanomag SS100—and then driven to the designated launching position.

Back at the railhead, the liquid oxygen (A-stoff) railway tanker was approached by members of the supply detachment. Hoses were attached from the railway tanker to the Betriebstoffanhanger (towed liquid oxygen bowser), which had been moved alongside the rail tanker. It was normally pulled by a heavy tractor such as the Hanomag SS100. The trailer carried a cylindrical bowser on top and was heavily reinforced to withstand the internal pressure of hundreds of gallons of liquid oxygen. Using a small gasoline-powered engine to run an external pump, the liquid oxygen bowser was filled; the super cooled liquid oxygen soon created frost on the hoses and attachments. Crewmen were required to wear special gloves to prevent injury when handling the extremely cold liquid oxygen. When the liquid oxygen bowser was full, it was towed to the fueling troop's parking area near the launching areas. Two of these bowsers could hold enough liquid oxygen to launch three rockets; about six bowsers could be filled by one railway tanker, enough to fire nine rockets.

The concentration of troops and vehicles preparing a V-2 for launch would be a great temptation to any Allied fighter looking for ground targets. So, the most ideal firing sites were those that provided heavy concealment, along with a flat, firm footing for the firing table. The firing crews were not limited to using prepared concrete launching pads, as they frequently used asphalt roads, paved ways, and even firm soil, which had been reinforced with timbers or steel sleepers. Clearings in wooded or forested areas were most ideal, as the tall pine and evergreen trees of Holland and Germany not only provided thick camouflage but also shielded the rockets from strong wind gusts, which could alter the rocket's course during its slow liftoff. The firing crews preferred to launch in the late evening hours near dusk.

The firing table, which was towed by the firing crew's Feuerleitpanzer (firing control vehicle) or Hanomag SS100 tractor, was pulled into position on the platform at the launching site. Once the legs of the firing table were extended and screwed down, the table was raised enough for the towing dolly to be removed. The 3,600-pound Pfaff-manufactured Abschussplattform (firing table) was made from welded steel and was, in itself, very ingenious. It was relatively small and portable (considering the size of the V-2) and could easily support the weight of the fully fueled 14-ton rocket. Next, the firing crew would adjust the dial-sight on the firing table and then by means of a winch pull the Meillerwagen backward up to the firing table. The extending supports were swung out and screwed down at the rear of the Meillerwagen for the needed support while raising the rocket. A small, gasoline-powered Volkswagen engine operated two hydraulic rams, which were controlled by one man to raise the lifting arm of the Meillerwagen into the vertical position, with the rocket centered over the table. The table was brought up to meet the rocket until the fins were seated in the appropriate position. After the V-2 was resting on the table, the clamps were released, and the Meillerwagen was rolled back about three feet from the launch table. The raised arm was then used as a servicing tower, with platforms extending from its framework for crew members' access near the midsection and control compartments of the raised rocket. On the ground, the survey crew was busy measuring to make sure the rocket was level.

Part of the equipment of each firing battery included a trio of Kine-Theodolites. The survey platoon was responsible for setting the rocket in the vertical position using the theodolites and practical application of principles in which bearings were taken from a fixed point on two points horizontally opposed across the longitudinal axis on the outer circumference of the rocket at its thickest part. The tip of the rocket was then brought in line with the bisector of the angle formed by the two bearings. The actual course as plotted by the theodolites was sent down to the batteries from the draftsmen of the computation section at Battalion headquarters.[1]

Newly manufactured Meillerwagen trailers of the V-2 battery. Attached to each trailer is the Hanomag SS100 Schwerer Radschlepper. This Wehrmacht heavy tractor, equipped with a six-cylinder diesel engine, was a multipurpose vehicle used mainly by V-2 troops in The Hague. The SS100 was a civilian tractor prior to the war but was pressed into service for the military because of its rugged design. It was used in towing the Vidalwagen, which transported the V-2 rocket from the railhead to the rocket field stores of the technical troop. It was also used to tow the Meillerwagen, which transported the V-2 rocket to the launch site and erected the rocket. Additionally, the Hanomag SS100 was used to tow other trailers or trailer pumps. (Deutsches Museum Munich)

After the protective engine jet covers were removed from the venturi in the combustion chamber and the fragile carbon-graphite exhaust rudders were carefully bolted in place, the V-2 was ready for fueling. The rocket had to be fueled in the vertical position because the fuel tanks inside of the rocket were designed to withstand the weight of the fuel only in the upright position. Once the rocket was raised on its firing table, the supply detachment or fueling troop moved quickly to the firing site location. A convoy of vehicles—two Opel Kfz. 385 B-stoff Kesselwagens (alcohol tanker trucks) towing a trailer pump, the Hanomag SS100 heavy tractor towing the liquid oxygen bowser, and the Opel Blitz T-stoff (hydrogen peroxide) tanker truck—would all approach the rocket in an organized fashion. A number of prefueling tests were performed, including a dry purge of the rocket's fuel tanks with compressed nitrogen to locate any leaks. The fueling trucks and trailers would then approach.

The launching crews first filled the V-2 with alcohol, then liquid oxygen. The Meillerwagen arm was designed with built-in plumbing for permanent delivery of alcohol and liquid oxygen when fueling. The hoses were connected to the Meillerwagen at the tanker and at the top of the V-2 fuel tanks. With the use of a towed trailer pump, alcohol fueling took about ten minutes. A few minutes later, the liquid oxygen tanker was pulled to the opposite side of the rocket. After alcohol fueling was finished, the Kesselwagen moved away from the rocket. The liquid oxygen was then pumped into the rocket, always no more than one hour before firing to prevent the internal valves from freezing. The hoses and clamps soon were covered in frost from the super cooled liquid oxygen. During the liquid oxygen fueling, the hydrogen peroxide was manually pumped into a premeasured container mounted to the Meillerwagen, which was emptied into the T-stoff tank by gravity. The Z-stoff (sodium permanganate) was kept heated to quicken the reaction with T-stoff when powering the steam turbine. The Z-stoff was then removed from its heater and emptied into the rocket manually. With fueling completed, the liquid oxygen tanker would then pull away. A technician would climb up to the midsection joint and adjust the tension created by eight tons of added fuel.

Electrical power for the V-2 was provided by ground sources when it rested on the launching platform and by batteries while in flight. The Germans used the Stromversorgungswagen (power supply truck) for the preliminary electrical power to the V-2 during the testing of the rocket's electrical systems, rather than using the rocket's onboard batteries. The two-ton Steyr model 2000 was converted for this role. Every preflight power and testing switch was channeled through the Steyr. This vehicle contained a Drehstromgenerator 6 KVA powered by a three-phase 6,000-watt Zundapp motor, which was hooked into the local civilian power grid. To supply the correct amperes and voltage to the rocket there was an induction motor connected to a dynamo; this converted the power to the required strength. Continuous DC current was supplied to the rocket through a cable leading to the rocket batteries, and it took about one hour to charge the batteries. Another cable carried AC current for heating the rocket's internal parts to counter the influence of the cold liquid oxygen during fueling. Onboard the Steyr, there were five cable reels and a small cable-laying mechanism. The Steyr also pulled a small trailer that contained eight more cable reels. This mass of cables was for power and testing between multiple vehicles of the unit. From the Steyr, cables ran to the Meillerwagen, to the firing table, to the Feuerleitpanzer, and to the testing truck. It also carried a cable that connected the testing truck to the cable mast of the firing table.

After fueling was completed, the arm of the Meillerwagen was lowered and it was pulled away. Vehicles that remained close to the actual firing platform were dug in and covered with camouflage netting; others vehicles were dispersed either to forward parking areas about 500 yards away or to rear lines about three to four miles away. The V-2 was then oriented to its exact firing position by using a dial-sight incorporated into the firing table. After a series of preflight tests, the electronics and gyros were set and armed. The firing troops then would install the electrical igniter into the combustion chamber.

At this point, ground personnel would take cover in slit-trenches that had been prepared earlier. The launch control officer and two-man crew entered the Feuerleitpanzer firing control vehicle, which was located about 300–400 feet away from the rocket, usually down in a protective trench that was dug when the site was prepared. It was from this armored vehicle that the final remote tests of the steering controls were completed.

The control officer asked the soldier at the board, "Steuerung klar?" (Firing control ready?)

The answer: "Steuerung klar!"

Everything was quiet. The soldiers in the trenches were only whispering.

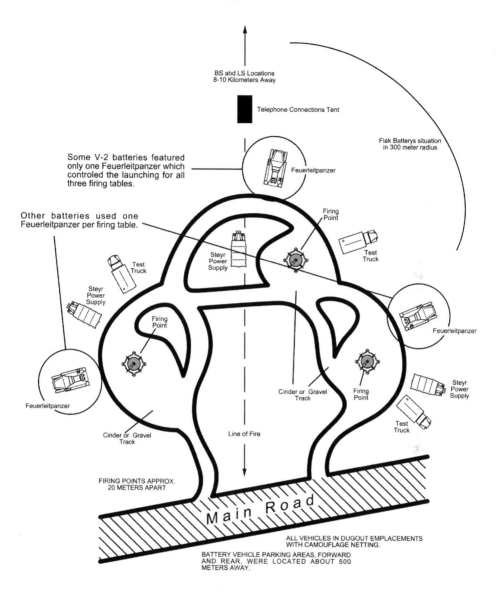

The launch officer called, "X-1" (T-minus one minute).

The officer stepped on a small perch in the Feuerleitpanzer. He was able to see the launch site. "Schlüssel auf Schiessen" (Turn the key for ignition), he ordered.

"Ist auf Schiessen, Klarlampe leuchtet!" (Ignition engaged, indicator light is on) said the man behind the controls.

The liquid oxygen and alcohol then flowed by gravity to the exhaust nozzle, where the igniter, which resembled a pinwheel that sparked as it rotated, commenced the burning. This gravity-flow burning generated only eight tons of thrust, which was not enough to launch the 14-ton rocket; however, it gave the firing control officer a visual indication that the flame below the rocket was burning properly. Once the firing control officer believed the rocket was ready to fire, an electric command was sent to start the fuel pumps; his last order could barely be heard over the roar of the engine.

"Hauptstufe!" (Main stage)

The soldier at the propulsion controls then pushed the button, and the fuel pumps began to scream. When the steam turbine reached top speed, the fuel flow was pushing at 275 pounds per second. The earth was shaking and vibrating under the pressure of some 60,000 pounds of thrust. The rocket went straight up and turned itself slowly to the target. After the rocket was airborne, the fuses in the warhead were armed automatically by a device known as the Sterg. The soldier at the propulsion board next turned the spanner of the high-pressure bottles down. The crews returned to the launch site and in only 30 minutes they would completely breakdown and withdraw, making it impossible for Allied aircraft to locate their position after the fact. At the end of the day, each firing crew would submit detailed reports about each event.

The V-2 would have a standard operational range of approximately 230 miles. Thirty seconds after launch, the rocket passed through the sound barrier. The maximum burning time of the engine was 65–70 seconds. Shortly before engine shutdown, at a height of approximately 22 miles, the missile weighed 8,900 pounds. After shutdown, the rocket flew to a height of 52 to 60 miles and fell to earth with an impact speed of approximately 2,100 miles per hour. At impact, in less than one-thousandth of a second, the fuse on top of the warhead ignited a small primary charge that ran up the middle of the warhead to ignite the main explosive, a mixture of TNT and ammonium nitrate (amatol). At exactly the same moment, in the base of the warhead, another two fuses were set off by an inertial switch. When launched against targets close to the operational range of the rocket, the deviation between target and impact was normally one to nine miles. This made the rocket only suitable for use against widely populated areas at that range. At shorter targeting ranges, the accuracy of the V-2 was improved.

A typical V-2 battery consisted of five primary sections, along with various other units. All battery personnel carried the usual combat weapons of normal Wehrmacht units, along with the specific equipment associated with their duties.

HEADQUARTERS SECTION: This section maintained the battery troop headquarters. It was responsible for general military administration and communication and also for range calculations and map plotting, as it was done within the battery.

LAUNCHING SECTION: The launching troops were split into three individual firing platoons, each with its own Feuerleitpanzer firing control vehicle and launching table. In most situations there were three Meillerwagens for each platoon; a total of nine Meillerwagens per battery. Each firing platoon contained around 39 men, all with a specific duty to perform during the execution of certain routine tests in the preparation of the rocket. Platoons included the fire control crew with the Feuerleitpanzer; the surveying and adjustment crew, which was responsible for the final survey of the launch site and for the actual alignment of the rocket before firing; the engine crew; the electrical crew; and the vehicle crew handling the Meillerwagen and firing table. Ordinarily, each firing platoon operated separately from the other platoons, although they sometimes fired simultaneously.

THE RADIO CONTROL SECTION: These soldiers manned the Brennschluss and Leitstrahl antenna arrays, each of which was carried in an Opel Blitz truck placed at a calculated distance of between eight and ten kilometers behind the firing point.

THE SUPPLY SECTION: This section consisted of four companies responsible for unloading the rockets and rocket fuels from the railhead. At the receiving station, the first company of the supply detachment would transport the missiles to the field store. They used about three Vidalwagen road transport trailers for each Meillerwagen, making a total of approximately 27 Vidalwagens per battery. The warheads were loaded onto trucks and brought separately to the field store. The second and third companies of the supply detachment, known as the fueling column, were responsible for the transport of rocket fuels from the railhead to the firing sites, while the fourth company supplied all vehicles of the battery with gasoline and oil.

THE TECHNICAL SECTION: These technicians manned the field stores and were responsible for testing, and if necessary, repairing any possible defects on the rockets. The technical troops would then prepare the rockets, with warheads attached, for the firing platoons.

Other various platoons included a maintenance section, which was responsible for maintenance of the unit's motorized transport vehicles and other mechanical equipment; a telephone section responsible for maintaining communications between the sections and with battalion headquarters; a fire fighting platoon; a camouflage and smoke section, which was responsible for building the camouflage arrangements around the firing positions; and a guard platoon to secure the areas into and around the firing sites. All of the V-2 positions were defended by designated flak platoons, which were supplemented by two heavy flak batteries at times.

V-2

Meillerwagen Erector Trailer

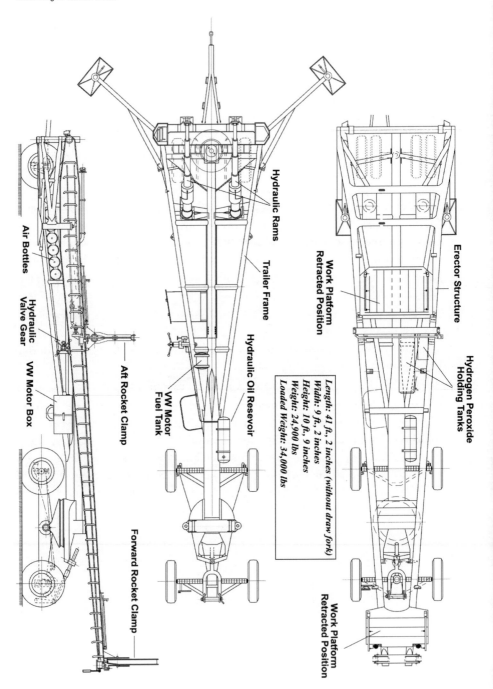

Hydraulic Rams

Trailer Frame

Work Platform
Retracted Position

Erector Structure

Hydrogen Peroxide
Holding Tanks

Air Bottles

Hydraulic
Valve Gear

VW Motor Box

Aft Rocket Clamp

VW Motor
Fuel Tank

Hydraulic Oil Resevoir

Forward Rocket Clamp

Work Platform
Retracted Position

Length: 41 ft. 2 inches (without draw fork)
Width: 9 ft. 2 inches
Height: 10 ft. 9 inches
Weight: 24,900 lbs
Loaded Weight: 34,000 lbs

V-2 Rocket

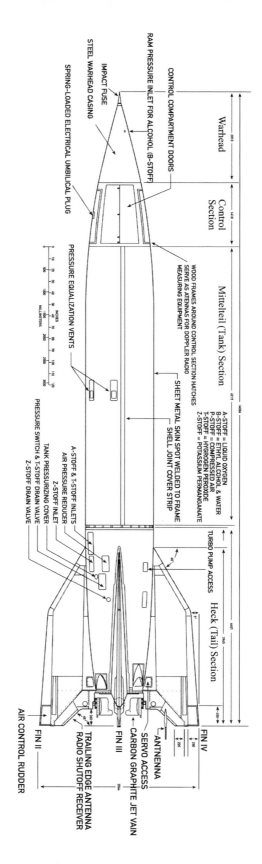

Warhead

RAM PRESSURE INLET FOR ALCOHOL (B-STOFF)
CONTROL COMPARTMENT DOORS
IMPACT FUSE
STEEL WARHEAD CASING
SPRING-LOADED ELECTRICAL UMBILICAL PLUG

Control
Section

Mittelteil (Tank) Section

WOOD FRAMES AROUND CONTROL SECTION HATCHES
SERVE AS ATENNAS FOR DOPPLER RADIO
MEASURING EQUIPMENT

PRESSURE EQUALIZATION VENTS

A-STOFF = LIQUID OXYGEN
B-STOFF = ETHYL ALCOHOL & WATER
P-STOFF = COMPRESSED AIR
T-STOFF = HYDROGEN PEROXIDE
Z-STOFF = POTASSIUM PERMANGANATE

SHEET METAL SKIN SPOT WELDED TO FRAME
SHELL JOINT COVER STRIP

A-STOFF & T-STOFF INLETS
AIR PRESSURE REDUCER
Z-STOFF INLET
TANK PRESSURIZING COVER
PRESSURE SWITCH & T-STOFF DRAIN VALVE
Z-STOFF DRAIN VALVE

TURBO PUMP ACCESS

Heck (Tail) Section

AIR CONTROL RUDDER
FIN II
TRAILING EDGE ANTENNA
RADIO SHUTOFF RECEIVER
FIN III
ANTNENNA
SERVO ACCESS
CARBON GRAPHITE JET VAIN
FIN IV

Notes

Chapter One

1. Dornberger, *V-2*
2. King and Kutta, *Impact*
3. Ibid.
4. Ordway and Sharpe, *The Rocket Team*
5. Bergaust, *Reaching for the Stars*
6. Garli?ski, *Hitler's Last Weapons*
7. De Maeseneer, *Peenemünde*
8. Ordway and Sharpe, *The Rocket Team*
9. De Maeseneer, *Peenemünde*
10. Ordway and Sharpe, *The Rocket Team*
11. Lampton, *Wernher von Braun*
12. De Maeseneer, *Peenemünde*
13. Dornberger, *V-2*
14. Neufeld, *The Rocket and the Reich*
15. Ordway and Sharpe, *The Rocket Team*
16. Dornberger, *V-2*
17. Neufeld, *The Rocket and the Reich*
18. Dornberger, *V-2*
19. Ibid.
20. Heitmann, *"The Peenemünde Rocket Center"*
21. Ordway and Sharpe, *The Rocket Team*
22. Neufeld, *The Rocket and the Reich*
23. King and Kutta, *Impact*
24. De Maeseneer, *Peenemünde*
25. Dornberger, *V-2*
26. Ramsey, *"The V-Weapons"*
27. Neufeld, *The Rocket and the Reich*
28. Dornberger, *V-2*
29. Jones, *Most Secret War*
30. Neufeld, *The Rocket and the Reich*
31. Ibid.

Chapter Two

1. De Maeseneer, Peenemünde
2. Speer, Inside the Third Reich
3. NASM FE342; Neufeld, The Rocket and the Reich
4. Neufeld, The Rocket and the Reich

5. NASM FE338; Hölsken, V-Missiles of the Third Reich

6. NASM FE341; Neufeld, The Rocket and the Reich

7. Baker, *The Rocket: The History and Development of the Rocket and Missile Technology*

8. Ordway and Sharpe, *The Rocket Team*

9. Speer, *Inside the Third Reich*

10. Baker, *The Rocket: The History and Development of Rocket & Missile Technology*

11. Stine, *ICBM: The Making of the Weapon That Changed the World*

12. De Maeseneer, *Peenemünde*

13. Kennedy, *Vengeance Weapon 2: The V-2 Guided Missile*

14. De Maeseneer, *Peenemünde*

15. Ordway and Sharpe, *The Rocket Team*

16. Konrad Dannenberg, personal account, 2001

17. Dornberger, *V-2*

18. Ordway and Sharpe, *The Rocket Team*

19. Neufeld, *The Rocket and the Reich*

20. Speer, *Inside the Third Reich*

21. Neufeld, *The Rocket and the Reich*

22. Ibid.

23. Ramsey, *"The V-Weapons"*

24. Hautefeuille, *Constructions Spéciales*

25. Dornberger, *V-2*

26. Ibid.; PRO WO208/4178

27. Ordway and Sharpe, *The Rocket Team*

Chapter Three

1. De Maeseneer, *Peenemünde*

2. Jones, *Most Secret War*

3. McGovern, *Crossbow and Overcast*

4. Casey, *"Surveillance in WWII"*

5. Jones, *Most Secret War*

6. At least one of the rockets observed in Prüfstand VII by Allied reconnaissance on June 23, 1943, may have been a complete 1:1 scale wooden mockup of a V-2 used for training purposes. It was painted white or possibly some other light color and would have been easily noticeable in the brilliant sunlight of that particular day.

7. Jones, *Most Secret War*

8. Baker, *The Rocket: The History and Development of Rocket & Missile Technology*

9. Ordway and Sharpe, *The Rocket Team*

10. Helfers, *The Employment of V-Weapons by the Germans During World War II*; King and Kutta, *Impact*

11. Allied aircrews reported a massive amount of antiaircraft batteries situated near the Eperlecques bunker.

12. Hautefeuille, *Constructions Spéciales*

13. Dornberger, *V-2*

14. Ibid.

15. Casey, *"Surveillance in WWII"*

16. McGovern, *Crossbow and Overcast*

17. Hölsken, *V-Missiles of the Third Reich*; Neufeld, *The Rocket and the Reich*

18. Speer, *Inside the Third Reich*; Dornberger, *V-2*

19. Padfield, *Himmler*

20. Sellier, *Histoire du camp de Dora*

21. Beon, *Planet Dora*; Grigorieff, *"Mittelbau GmbH - Mittelbau KZ"*

22. Michel, *Dora*; Grigorieff, *"Mittelbau GmbH - Mittelbau KZ"*

23. PRO AIR 8/1368, Crossbow Cherbourg Peninsular Sites, July 1944; PRO AIR 40/1777, Crossbow Report on Flying Bomb and Rocket Firing and Storage Sites, August 1943 October 1944

24. Casey, *"Surveillance in WWII"*

25. Ramsey, *"An Engineer Returns"*; Ramsey, *"The V-Weapons"*

26. Neufeld, *The Rocket and the Reich*

27. Collier, *Battle of the V-Weapons 1944–1945*

28. Casey, *"Surveillance in WWII"*

29. Gruen, Preemptive Defense, *Allied Air Power Versus Hitler's V-Weapons 1943–1945* United States Strategic Bombing Survey, The Air Offensive Against V-Weapons #60, 1945 Narrative of Operations., Mission 164, December 24, 1943; Joseph W. Angell, Chapter 4: Crossbow, Wesley Frank Craven and James Lea Cate, The Army Air Forces in World War 2, III, Europe: Argument to V-E Day January 1944 to May 1945; New York Times, December 24, 1943

30. Dwight D. Eisenhower, *Crusade in Europe*

31. Helfers, *The Employment of V-Weapons by the Germans During World War II*

32. Ibid.

33. Ibid.

34. Kennedy, *Vengeance Weapon 2: The V-2 Guided Missile*; Dornberger, *V-2*

35. Hölsken, *V-Missiles of the Third Reich*

Chapter Four

1. Glass, Kordaczuk, St?pniewska, *Wywiad Armii Krajowej W Walce Z V-1 / V-2*

2. Hölsken, *V-Missiles of the Third Reich*

3. Ordway and Sharpe, *The Rocket Team*

4. A.D.I. Report No. 633/1944

5. Ibid.

6. Ibid.; Dornberger, *V-2*; King and Kutta, *Impact*

7. Sellier, *Histoire du camp de Dora*

8. Wagner, *Produktion des Todes - Das KZ Mittelbau Dora*

9. Michel, *Dora*; Grigorieff, *"Mittelbau GmbH - Mittelbau KZ"*

10. De Maeseneer, *Peenemünde*

11. Ibid.; Ordway and Sharpe, *The Rocket Team*

12. King and Kutta, *Impact*

13. Cooksley, *Flying Bomb*

14. Craven and Cate, *The Army Air Forces in World War II*

15. United States Strategic Bombing Survey, Report No. 60, The Air Offensive Against V-Weapons, 1945

16. Neufeld, *The Rocket and the Reich*

17. Ordway and Sharpe, *The Rocket Team*

18. Ibid.

19. Dornberger, *V-2*; Speer, *Inside the Third Reich*; Neufeld, *The Rocket and the Reich*

20. Dornberger, *V-2*

21. Lampton, *Wernher von Braun*

22. Dornberger, *V-2*; Cor Lulof, 2001

23. Dornberger, *V-2*

24. Helfers, *The Employment of V-Weapons by the Germans During World War II*

25. Ibid.

26. Dornberger, *V-2*; Neufeld, *The Rocket and the Reich*

27. A.D.I. Report No. 633/1944

28. De Maeseneer, *Peenemünde*

29. Jones, *Most Secret War*

30. PRO WO.208/4178

31. Hölsken, *V-Missiles of the Third Reich*; Neufeld, *The Rocket and the Reich*

32. Neufeld, *The Rocket and the Reich*; Dornberger, *V-2*

33. Hölsken, *V-Missiles of the Third Reich*

34. Heinz Wellmann, personal account, former soldier of the third platoon of SS Werfer Abteilung 500, courtesy of Dr. Karl-Heinz Wellmann, 2004

35. A.D.I. Report No. 633/1944; Dornberger, *V-2*

36. Dornberger, *V-2*

37. De Maeseneer, *Peenemünde*; United States Strategic Bombing Survey, Report No. 60, The Air Offensive Against V-Weapons, 1945

38. Jacek Kruk, 2004; Detlev Paul, 2004

39. Hölsken, *V-Missiles of the Third Reich*; United States Strategic Bombing Survey, Report No. 60, The Air Offensive Against V-Weapons, 1945

40. Longmate, *Hitler's Rockets: The Story of the V-2s*

41. A.D.I. Report No. 633/1944; PRO WO.208/3121; Neufeld, *The Rocket and the Reich*

42. Neufeld, *The Rocket and the Reich*

43. Ibid.; Dornberger, *V-2*

44. Hölsken, *V-Missiles of the Third Reich*; Helfers, *The Employment of V-Weapons by the Germans During World War II*; BAMA RH41.1195, War Diary of Battalion 836

45. Helfers, *The Employment of V-Weapons by the Germans During World War II*; Grailet, *Première Mondiale Pour le V-2 sur Paris*

46. Hölsken, *V-Missiles of the Third Reich*; Helfers, *The Employment of V-Weapons by the Germans During World War II*

47. Johnson, *V-1 V-2: Hitler's Vengeance on London*

Chapter Five

1. BAMA RH 26.1022/3, Division z.V. War Diary; Ramsey, *"The V-Weapons"*

2. De Maeseneer, *Peenemünde*; Johnson, *V-1 V-2: Hitler's Vengeance on London*

3. Jones, *Most Secret War*

4. Ogley, *Doodlebugs and Rockets*; Johnson, *V-1 V-2: Hitler's Vengeance on London*

5. Hölsken, *V-Missiles of the Third Reich*

6. United States Strategic Bombing Survey, Report No. 20, V Weapons in London, 1947

7. Johnson, *V-1 V-2: Hitler's Vengeance on London*

8. Helfers, *The Employment of V-Weapons by the Germans During World War II*; BAMA RH 26.1022/3, Division z.V. War Diary

9. PRO WO.208/3121

10. Longmate, *Hitler's Rockets: The Story of the V-2s*

11. Ibid.; Ogley, *Doodlebugs and Rockets*

12. BAMA RH 26.1022/3, Division z.V. War Diary

13. Cor Lulof, 2004

14. PRO WO.208/3121; Heinz Wellmann, personal account, former soldier of the third platoon of SS Werfer Abteilung 500, courtesy of Dr. Karl-Heinz Wellmann, 2004; Detlev Paul, 2002

15. Henk Koopman, 2004

16. Jungbluth, *Hitler's Geheimwaffen im Westerwald*

17. A.D.I. Report No. 633/1944; Detlev Paul, 2002

18. Detlev Paul, 2002; BAMA RH 26.1022/3, Division z.V. War Diary; Hölsken, *V-Missiles of the Third Reich*

19. Official "Big Ben" Mission Report, Colonel T.R.B. Sanders, 1944

20. Balk City Hall Records

21. Ogley, *Doodlebugs and Rockets*; Collis and Ramsey, *The Blitz Then and Now Volume 3*

22. Henk Koopman, 2001

23. Jan Willem Draijer, 2004

24. Collis and Ramsey, *The Blitz Then and Now Volume 3*

25. Cabell and Thomas, *Operation Big Ben*

Chapter Six

1. Rita Koenen-Winter, personal account, 2002

2. Johnson, *V-1 V-2: Hitler's Vengeance on London*

3. Henk Koopman, 2005; Rita Koenen-Winter, personal account, 2002

4. Hölsken, *V-Missiles of the Third Reich*

5. Porezag, *Geheime Kommandofache*; BAMA RH41.1192, 1195, 1198, War Diary Battalion 836

6. Johnson, *V-1 V-2: Hitler's Vengeance on London*

7. Beon, *Planet Dora*; Grigorieff, *"Mittelbau GmbH - Mittelbau KZ"*

8. Speer, *Inside the Third Reich*

9. Dornberger, *V-2*; De Maeseneer, *Peenemünde*

10. Ogley, *Doodlebugs and Rockets*; Cabell and Thomas, *Operation Big Ben*

11. David Johnson, V-1 V-2: Hitler's Vengeance on London, 1982

12. Cabell and Thomas, *Operation Big Ben*

13. Ogley, *Doodlebugs and Rockets*

14. Johnson, *V-1 V-2: Hitler's Vengeance on London*

15. United States Strategic Bombing Survey, Report No. 20, V Weapons in London, 1947

16. BAMA RH 26.1022/3, Division z.V. War Diary; Heinz Wellmann, personal account, former soldier of the third platoon of SS Werfer Abteilung 500, courtesy of Dr. Karl-Heinz Wellmann, 2004; Kooy, *Ballistics of the Future*

17. Cor Lulof, personal account, 2001

18. Hindriks and Spoor, *Wernher von Braun; één van de 118*

19. Heinz Wellmann, personal account, former soldier of the third platoon of SS Werfer 20. Abteilung 500, courtesy of Dr. Karl-Heinz Wellmann, 2004

20. Porezag, *Geheime Kommandofache*

21. Detlev Paul, 2002; Claudio Becker, 2002

22. Gruen, *Preemptive Defense: Allied Air Power Versus Hitler's V-Weapons 1943–1945*

23. BAMA RH 26.1022/3, Division z.V. War Diary; Ordway and Sharpe, *The Rocket Team* This event has been mistakenly reported as having occurred at a castle named "Schloss Burg" located in the Ruhr Valley near Remscheid. However, there were no V-2 troops launching rockets near Remscheid at any point during the war.

Chapter Seven

1. Kenneth Hartman, personal account, 2003

2. Franssen, *De Slag om Antwerpen*; De Schuyter, *Gemarteld Antwerpen*

3. De Maeseneer, *Peenemünde*

4. Greg Hayward, personal account, 2002

5. Kenneth Hartman, personal account, 2003

6. Rely, *"City of Sudden Death"*

7. Simone De Ceunynck, personal account, 2003

8. Lynne Robinson-Lawrence, 2004

9. Charles Ostyn, personal account, 2001

10. Palinckx, *Antwerpen Onder De V-bommen 1944-1945*

11. James Mathieson, personal account, 2005

12. De Schuyter, *Gemarteld Antwerpen*; Charles Ostyn, personal account, 2001

13. Franssen, *De Slag om Antwerpen*

14. De Schuyter, *Gemarteld Antwerpen*

15. Achiel Rely, *"City of Sudden Death"*; King and Kutta, *Impact*; Charles Ostyn, personal account, 2001

16. King and Kutta, *Impact*; PRO WO.205/999, Defense of Antwerp and Belgium

17. Ambrose, revised from Sulzberger, *New History of World War II*

18. King and Kutta, *Impact*

Chapter Eight

1. King and Kutta, *Impact*

2. Dornberger, *V-2*; Emme, *Aeronautics and Astronautics: An American Chronology of Science and Technology in the Exploration of Space 1915-1960*

3. Dornberger, *V-2*

4. Michels, *Peenemünde und seine Erben in Ost und West*

5. Simpson, *The Greatest Squadron of Them All: The Definitive History of 603 Squadron*

6. Cabell and Thomas, *Operation Big Ben*; Simpson, *The Greatest Squadron of Them All: The Definitive History of 603 Squadron*

7. Collier, *The Battle of the V-Weapons 1944-45*

8. Huurman, *Het Spoorwegbedrijf in Oorlogstijd 1940 – 1945*

9. Cabell and Thomas, *Operation Big Ben*

10. Bill Wils, personal account, 2000

11. Cor Lulof, personal account, 2001

12. Cornelissen, *Van grasmat tot Fliegerhorst: een oorlogsdocumentatie over de luchtstrijd boven het oostelijk deel van Overijssel / Oldenzaal: Twents-Gelderse Uitgeverij De Bruyn*

13. Heinz Wellmann, personal account, former soldier of the third platoon of SS Werfer Abteilung 500, courtesy of Dr. Karl-Heinz Wellmann, 2004

14. Ibid.

15. Cor Lulof, personal account, 2004

16. Rita Koenen-Winter, personal account, 2002

17. Ibid.

18. Hechler, *The Bridge at Remagen*

19. PRO WO.208/3121; Hölsken, *V-Missiles of the Third Reich*

20. PRO WO.291/287, RAF Type 9 Mark V Radar Tracking, 1945

21. Hechler, *The Bridge at Remagen*; BAMA RS4.1542, War Diary SS Werfer Battery 500; PRO WO 219/4932

22. PRO HW.1/3618

23. Dornberger, *V-2*; PRO WO.219/4937

24. PRO WO.208/3121

25. Ordway and Sharpe, *The Rocket Team*

26. War Office, London, Report on Operation "Backfire", 1946

27. Collier, *The Battle of the V-Weapons 1944–45*

28. Heinz Wellmann, personal account, former soldier of the third platoon of SS Werfer Abteilung 500, courtesy of Dr. Karl-Heinz Wellmann, 2004

29. Michael Grube, 2002; Michael Keuer, 2003

30. Hölsken, *V-Missiles of the Third Reich*

31. Heinz Wellmann, personal account, former soldier of the third platoon of SS Werfer Abteilung 500, courtesy of Dr. Karl-Heinz Wellmann, 2004; Russell, *No Triumphant Procession*

32. PRO WO.208/3153, Surrender of the V-2 Division, SS Lieutenant Colonel Wolfgang Wetzling, 1945.

33. Heinz Wellmann, personal account, former soldier of the third platoon of SS Werfer Abteilung 500, courtesy of Dr. Karl-Heinz Wellmann, 2004

34. Sellier, *Histoire du camp de Dora*; Grigorieff, *"Mittelbau GmbH - Mittelbau KZ"*

35. PRO WO.208/4178

36. McGovern, *Crossbow & Overcast*

37. Agoston, *Blunder!*

Chapter Nine

1. Neufeld, *The Rocket and the Reich*; Grigorieff, *"Mittelbau GmbH - Mittelbau KZ"*

2. Grigorieff, *"Mittelbau GmbH - Mittelbau KZ"*

3. Neufeld, *The Rocket and the Reich*

4. McGovern, *Crossbow and Overcast*; Grigorieff, *"Mittelbau GmbH - Mittelbau KZ"*

5. Reuter, *The V-2 and the Russian and American Rocket Program*

6. Ramsey, *"The V-Weapons"*

7. War Office, London, Report on Operation "Backfire", 1946

8. McGovern, *Crossbow & Overcast*

9. Ibid.

10. PRO WO.208/4178

11. Neufeld, *The Rocket and the Reich*; Ordway and Sharpe, *The Rocket Team*

12. Godwin, *Dyna-Soar: Hypersonic Strategic Weapons System*

13. King and Kutta, *Impact*; Gruen, *Preemptive Defense: Allied Air Power Versus Hitler's V-Weapons 1943–1945*; Major Bunn Hearn, Headquarters, U.S. War Department, Memorandum, Subject: V-Weapon Activity Daily Report, 1945

14. Dwight D. Eisenhower, *Crusade in Europe*

15. McDougall, *The Heavens and the Earth: A Political History of the Space Age*

16. Ross, *Strategic Bombing by the United States in World War II: The Myths and the Facts*

17. United States Strategic Bombing Survey, The Air Offensive Against V-Weapons #60, 1945; King and Kutta, *Impact*

18. Gruen, *Preemptive Defense, Allied Air Power Versus Hitler's V-Weapons 1943–1945*

19. Casey, *"Surveillance in WWII"*

Appendix

1. A.D.I. Report No. 633/1944

Sources and Bibliography

Archive Source Documents
Public Record Office, London (PRO)
AIR.14/2753, 3727
AIR.20/2629
AIR.34/80
AIR.37/1253
AIR.40/1777
AIR.51/425
AVIA.7/2395
CAB.119/7
HW.1/3618
WO.208/3121, 3153, 4178
WO.219/708, 4937, 5291
WO.291/287
A.D.I. Report No. 306/1945
A.D.I. Report No. 633/1944

National Air and Space Museum (NASM)
FE/303, 331, 338, 342, 733, 852

Bundesarchiv/Militärarchiv Freiburg (BAMA)
RH12.14/1
RH26.1022/3, 1022/6
RH41.1192, 1195, 1198, 1199, 1200
RS4.1542

Interviews and Correspondence
Balk, Netherlands. City Hall Records
Becker, Claudio. Correspondence with author, 2002.
De Ceunynck, Simone. Personal account. Interview by correspondence, 2003.
Draijer, Jan Willem. Correspondence with author, 2004.
Grube, Michael. Correspondence with author, 2002.
Hartman, Kenneth. Personal account. Interview by correspondence, 2003–2004
Hayward, Greg. Personal account. Interview by correspondence, 2002.
Keuer, Michael. Correspondence with author, 2003.
Koenen-Winter, Rita. Personal account. Interview by correspondence, 2002.
Koopman, Henk. Correspondence with author, 2001–2005.
Lulof, Cor. Personal account. Interview by correspondence, 2001–2005.
Mathieson, James. Personal account. Interview by correspondence, 2005.
Ostyn, Charles. Personal account. Interview by correspondence, 2001–2004.

Paul, Detlev. Correspondence with author, 2002–2005.

Robinson-Lawrence, Lynne. Correspondence with author, 2004.

Wellmann, Heinz. Personal account (former soldier of the third platoon of SS Werfer Abteilung 500). Interview by correspondence, 2004. Courtesy of Dr. Karl-Heinz Wellmann.

Wils, Bill. Personal account. Interview by correspondence, 2000.

Reference Books, Articles and Reports

AC/AS Intelligence Memorandum. German Pilotless Aircraft. London: Crossbow Reports, June 1944.

Agoston, Tom. BLUNDER! How the U.S. Gave Away Nazi Supersecrets to Russia. London: William Kimber, 1986.

Ambrose, Stephen E. D-Day June 6, 1944: The Climactic Battle of World War II. New York: Simon and Schuster, 1994.

Ambrose, Stephen E., revised from C.L. Sulzberger. New History of World War II. New York: Viking Press, 1997.

Babington-Smith, Constance. Evidence in Camera. London: Chatto and Windus, 1958.

Baker, David, Ph.D. The Rocket: The History and Development of Rocket and Missile Technology. London: Crown Publishers, 1979.

Bergaust, E. Reaching for the Stars. New York: Doubleday, 1960.

Béon, Yves. Planet Dora. Colorado: Westview Press, 1997.

Official "Big Ben" Mission Report, Colonel T.R.B. Sanders, 1944

Boer, J. F. A. Raketten Over Den Haag. Amsterdam: Die Haeghe Jaarboek, 1948.

Bowman, H. L. "V Weapons in London." United States Strategic Bombing Survey, Report No. 20 (1947).

Bullock, Allen. Hitler: A Study in Tyranny. New York: Harper and Rowe Publishers, 1962.

Cabell, Craig, and Graham A Thomas. Operation Big Ben: The Anti-V2 Spitfire Missions 1944–1945. Kent: T. J. International Ltd., 2004.

Carter, Col. W. S. J. (Air Defense Division, Supreme Headquarters, Allied Expeditionary Forces Memorandum). BIG BEN Attack on the Remagen Bridgehead. Versailles: SHAEF, April 1945.

Casey, Dennis. Surveillance in WWII. Lackland AFB, Texas: Air Intelligence Agency.

Cels, Jos. V-Bommen op Antwerpen (V-Bombs on Antwerp). Antwerp: Standaard Uitgeverij, 1984.

Churchill, Winston S. The Second World War: Triumph and Tragedy. Boston: Houghton Mifflin Company, 1953.

Collier, Basil. The Battle of the V-Weapons. London: Hoddler and Stoughton, 1964.

Collis, Robert and Winston G. Ramsey. The Blitz Then and Now Volume 3. London: Battle of Britain Prints International Limited, 1990.

Cooksley, Peter G. Flying Bomb. New York: Charles Scribner's Sons, 1979.

Cornelissen, C. Van grasmat tot Fliegerhorst: een oorlogsdocumentatie over de luchtstrijd boven het oostelijk deel van Overijssel. Oldenzaal: Twents-Gelderse Uitgeverij De Bruyn, 1998.

Craven, Wesley Frank, and James Lea Cate. The Army Air Forces in World War 2, Vol. III, Europe: Argument to V-E Day January 1944 to May 1945. Chicago: University of Chicago Press, 1951.

Cuich, Myrone N. Armes Secrètes et Ouvrages Mystérieux, les V1 et V2, volume 2. Tourcoing: édité à compte d'auteur, 1996.

Culver, Bruce. "The A4/V2 Missile." Allied-Axis Photo Journal No. 1 (2000): pp. 2–27.

De Maeseneer, Guido. Peenemünde: The Extraordinary Story of Hitler's Secret Weapons V-1 and V-2. Vancouver: AJ Publishing, 2001.

De Schuyter, Jan. Gemarteld Antwerpen: sinjorenstad onder de V Bommen; met teek. door Gaston Schuermans. Antwerpen: De Palm, 1945.

Dornberger, Walter. "Can Russian Missiles Strike the United States?" Collier's Magazine (January 1955).

Dornberger, Walter. V-2. Translated by James Cleugh and Geoffrey Halliday. New York: Viking Press, 1954.

Eisenhower, Dwight D. Crusade in Europe. Garden City, New York: Double Day and Company, Inc., 1952.

Emme, Eugene M. Aeronautics and Astronautics: An American Chronology of Science and Technology in the Exploration of Space 1915-1960. Washington DC: National Aeronautics and Space Administration, 1961.

Garli?ski, Józef. Hitler's Last Weapons. London: Julienn Friedmann, 1978.

Glass, A., S. Kordaczuk, and D. St?pniewska. Wywiad Armii Krajowej W Walce Z V-1 / V-2 (Polish Home Army Intelligence Service Against the V-1 and V-2). Warszawa: Mirage Hobby, 2000.

Godwin, Robert. Dyna-Soar: Hypersonic Strategic Weapons System. Burlington ON: Apogee, 2003.

Grailet, Lambert. Première Mondiale Pour le V-2 sur Paris. Tongeren: George Michiels, 1996

Grigorieff, Paul. Mittelbau GmbH - Mittelbau KZ, 2001.

Gruen, Adam L. Preemptive Defense, Allied Air Power Versus Hitler's V-Weapons 1943–1945. United States Air Force Historical Studies Office, 1991.

Gückelhorn, Wolfgang, and Detlev Paul. V1 "Eifelschreck," Abschüsse, Abstürze und Einschläge der fliegenden Bombs aus der Eifel und dem Rechtsrheinischen 1944/45. Aachen: Helios Verlags- und Buchvertriebsgesellschaft, 2004.

Hahn, Fritz. Deutsche Geheimwaffen. Heidenheim: Erich Hoffman Verlag, 1963.

Hallion, Richard P. Storm Over Iraq: Air Power and the Gulf War. Washington, D.C.: Smithsonian Institution Press, 1992.

Hautefeuille, Roland. Constructions Spéciales. Paris: Selbstverlag, 1985.

Hazenberg, F. R., A. N. W. Kenens, and R. van Lith. Wassenaar in de Tweede Wereldoorlog. Wassenaar: Stichting Wassenaar, 1995.

Hearn, Major Bunn (Headquarters, U.S. War Department Memorandum). V-Weapon Activity Report. April 7–8, 1945.

Hechler, Ken. The Bridge at Remagen: The Amazing Story of March 7, 1945—The Day the Rhine River was Crossed. 1957. Reprint, Missoula, Montana: Pictorial Histories Publishing Company, 1993.

Heitmann, Jan. "The Peenemünde Rocket Centre." After the Battle Magazine, no. 74 (1991): pp. 1–25.

Helfers, Lieutenant Colonel M.C. The Employment of V-Weapons by the Germans During World War II. Washington D.C.: Department of the Army, Office of the Chief of Military

History, 1954.

Hindriks, Klaas Jan and Rudolf Spoor. Wernher von Braun; één van de 118. Amsterdam: 1977.

Hölsken, Dieter. V-Missiles of the Third Reich: the V-1 and V-2. Sturbridge, Massachusetts: Monogram Aviation Publications, 1994.

Howard, Michael. British Intelligence in the Second World War. New York: Cambridge University Press, 1990.

Huurman, C. Het spoorwegbedrijf in oorlogstijd, 1939-1945. 's-Hertogenbosch: Uquilair, 2001.

Huzel, Dieter K. Peenemünde to Canaveral. New Jersey: Prentice-Hall, 1962.

Johnson, David. V-1 V-2: Hitler's Vengeance of London. New York: Stein and Day Publishers, 1981.

Jones, R. V. Most Secret War. London: Hamish Hamilton, 1978.

Jungbluth, Uli. Wunderwaffen im KZ "Rebstock," Zwangsarbeit in Dernau/Rheinland-Pfalz und Artern/Thüringen im Dienste der V-Waffen. Briedel/Mosel: Rhein-Mosel-Verlag, 2000.

Jungbluth, Uli. V1,2,3: Hitler's Geheimwaffen im Westerwald. Montabaur/Westerburg: Werkstatt-Beiträge zum Westerwald, 1996.

Kennedy Gregory P. Vengeance Weapon 2: The V-2 Guided Missile. Washington, D.C.: Smithsonian Institution Press, 1983.

King, Benjamin, and Timothy J. Kutta. Impact: The History of Germany's V-Weapons in World War II. New York: Sarpedon, 1998.

Klee, Ernst, and Otto Merk. Damals in Peenemünde. Oldenburg: Gerhard Stalling Verlag, 1963.

Kooy, J. M. J., and Uytenbogaart. Ballistics of the Future. New York: McGraw-Hill, 1946.

Lampton, Christopher. Wernher Von Braun. New York: Franklin Watts, 1988.

Lehman, Milton. Robert H. Goddard. New York: Da Capo Press, 1988.

Ley, Willy. Rockets, Missiles, and Men in Space. New York: Viking Press, 1968.

Longmate, Norman. Hitler's Rockets: The Story of the V-2s. London: Hutchinson Publishing, 1985.

Margry, Karel. "Nordhausen." After the Battle Magazine, no. 101 (1998): pp. 2–43.

McDougall, Walter A. The Heavens and the Earth: A Political History of the Space Age. New York: The Johns Hopkins University Press, 1997.

McElheran, Brock. V-Bombs and Weathermaps. London: McGill-Queen's University Press, 1995.

McGovern, James. Crossbow and Overcast. New York: William Morrow and Co., 1964.

Michel, Jean. Dora. Paris: Jean-Clade Lattäs, 1975.

Michels, Jürgen. Peenemünde und seine Erben in Ost und West, Entwicklung und Weg deutscher Geheimwaffen. Bonn: Bernard and Graefe Verlag, 1997.

Middlebrook, Martin. The Peenemünde Raid. London: Penguin Books, Ltd., 1982.

Neufeld, Michael J. "The Reichswehr, the Rocket, and the Versailles Treaty: A Popular Myth Reexamined." Journal of the BIS, Vol. 53 (2000): pp. 163–172.

Neufeld, Michael J. The Rocket and the Reich. New York: The Free Press, 1995.

Ogley, Bob. Doodlebugs and Rockets. Kent: Froglets Publications, 1992.

Ordway, Frederick I., III, and Mitchell Sharpe. The Rocket Team. New York: Thomas Y. Crowell, 1979.

Padfield, Peter. Himmler. New York: MJF Books, 1990.

Palinckx, Koen. Antwerpen Onder De V-Bommen 1944–1945. Antwerpen: Pandora, 2004.

Porezag, Karsten. Geheime Kommandosache: Geschichte der "V-Waffen" und geheimen Militäraktionen des Zweiten Weltkrieges an Lahn. Dill und im Westerwald: Verlag Wetzlardruck GmbH, Wetzlar, 1996.

Raiber, R, M.D. "The Führerhauptquartiere." After the Battle Magazine, no. 19 (1977): pp. 28–33.

Ramsey, Winston G., and Robert Collis. The Blitz Then and Now, Vol. 3. London: Battle of Britain Prints International, Ltd., 1990.

Ramsey, Winston G. "The V-Weapons." After the Battle Magazine, no. 6 (1974): pp. 2–37.

Rely, Achiel, and Winston Ramsey. "Antwerp: City of Sudden Death." After the Battle Magazine, no. 57 (1987): pp. 43–48.

Reuter, Claus. The V2 and the Russian and American Rocket Program. New York: S. R. Research and Publishing, 2000.

Ross, David M. S., Squadron Leader J. Bruce Blanche, William Simpson. The Greatest Squadron of Them All: The Definitive History of 603 Squadron, Volume Two: 1941 To Date. London: Grub Street, 2003.

Ross, Stewart Halsey. Strategic Bombing by the United States in World War II: The Myths and the Facts. Jefferson. NC: McFarland and Company, Inc., 2003.

Russell, John, with R. De Norman. No Triumphant Procession: The Forgotten Battles of April 1945. London: Arms and Armour, 1994.

Scheufelen, Klaus H. Mythos Raketen: Chancen für den Frieden. Munich/Esslingen: Bechtle Verlag, 2004.

Sellier, André. Histoire du Camp de Dora. Paris: La Découverte, 1998.

Sellier, André, and Ives Le Maner. Images De Dora 1943–1945. Centre d'Histoire de la Guerre et des Fusées, Saint-Omer, 1999.

Speer, Albert. Inside the Third Reich. New York: Avon, 1970.

Stine, G. Harry. ICBM: The Making of the Weapon that Changed the World. New York: Orion Books, 1991.

United States Army Ordnance. The Story of Peenemünde: What Might Have Been. Washington D.C.: United States War Department, 1945.

Von Braun, Wernher, and Frederick I. Ordway, III. History of Rocketry and Space Travel. London: Thomas Nelson and Sons, Ltd., 1967.

Von Braun, Wernher, and Frederick I. Ordway, III. The Rockets' Red Glare. New York: Doubleday, 1976.

Wagner, Jens-Christian. Produktion des Todes - Das KZ Mittelbau Dora. Göttingen: Wallstein Verlag, 2001.

The War Office. Report on Operation "Backfire." 5 Vols. London: January 1946.

Williams, Major C. R. "V-Weapons: Crossbow Campaign." U.S. Strategic Bombing Survey, Report No. 60 (1945).

Zeiler, Albert. Translation A4 Fibel (A4 Handbook, Peenemünde 1944). Huntsville, Alabama: Army Ballistic Missile Agency, 1957.

Index

A-1 rocket, 11

A-2 rocket, 11, 12, 16, 39

A-3 rocket, 12, 14, 16, 17, 18, 19, 22, 24

A-4 Fibel, 85

A-4 rocket, 2, 14, 15, 16, 18, 21, 22, 24, 25, 26, 27, 29, 32, 33, 34, 35, 36, 39, 40, 43, 44, 45, 46, 47, 48, 49, 50, 52, 53, 54, 55, 57, 62, 85, 133, 173, 201

A-4b rocket, 173

A-5 rocket, 18, 19, 21, 22, 24, 25, 27

Aachen, 17, 18

Aberdeen Proving Ground, 207, 212

Air Ministry (Luftwaffe), 14

Airburst, 92, 87, 98, 109, 121, 138, 146, 190

Altenwalde, 208, 209

American Interplanetary Society, 6

Antwerp, 2, 114, 122, 125, 134, 135, 139, 144, 148, 154, 156, 157, 159, 160, 161, 162, 163, 164, 165, 166, 167, 168, 169, 170, 171, 172, 174, 181, 182, 185, 188, 190, 194, 198, 203, 206, 215

Apollo, 20, 213

Army Weapons and Ordnance, 7, 8, 11, 12, 14, 16, 19, 29, 45, 48, 74, 97, 112

Archem, 146, 147, 150, 183

Ardennes, 115, 157, 166, 168, 171, 172, 173

Armsen, 196

Armstrong, C. H., 170

Arnhem, 123, 124, 215

Arnold, Henry H. "Hap", 92, 93

Austria, 18, 19, 49, 50, 71, 72, 104, 201, 203

Bäckebo, 105

Backfire, Operation, 208, 210, 212

Baltic, 13, 25, 39, 40, 43, 44, 45, 57, 59, 63, 66, 67, 109, 182, 196, 210

Barbarossa, 36

Battalion 485, 84, 85, 98, 110, 112, 113, 116, 118, 119, 120, 121, 123, 124, 125, 134, 135, 136, 138, 139, 159, 160, 161, 173, 174, 176, 180, 181, 194, 197

Battalion 836, 84, 85, 98, 99, 110, 112, 113, 126, 138, 139, 144, 156, 161, 162, 173, 174, 187, 188, 196, 197

Battery 444, 65, 79, 84, 85, 91, 108, 112, 113, 114, 115, 120, 121, 122, 128, 129, 132, 133, 134, 136, 137, 138, 139, 142, 148, 161, 173, 174, 196, 197

Battle of Britain, 33, 217

Battle of the Bulge, 157, 171, 172

Baumholder, 99, 113, 127

Becker, Karl, 7, 9, 10, 11, 14, 19, 20, 22, 29

Berg en Dal, 113, 124

Berlin, 4, 5, 6, 7, 8, 9, 10, 13, 28, 33, 35, 47, 48, 49, 50, 52, 62, 63, 66, 67, 69, 71, 95, 104, 105, 106, 107, 112, 122, 139, 195, 196, 198, 201, 202

Beukenhorst, 119, 123, 137

Bezuidenhout, 180, 186

Big Ben, 118, 127, 134

Bleicherode, 90, 193, 198

Blizna, 70, 83, 84, 86, 96, 97, 98, 99, 100, 108, 109, 127, 128

Blockhaus, 47, 72, 74, 76, 100, 106

Bloemendaal, 135, 142

Bodenplatte, Operation, 172

Borkum, 11

Bormann, Martin, 19

Brauchitsch, Walther von, 19, 22, 26, 27, 33, 34, 35, 36, 58

Braun, Magnus von, 95, 201, 202

Braun, Wernher von, 1, 4, 6, 7, 8, 9, 10, 12, 13, 14, 15, 16, 17, 20, 21, 24, 26, 28, 34, 36, 41, 43, 44, 48, 51, 52, 53, 54, 55, 56, 66, 68, 81, 90, 94, 95, 96, 158, 192, 193, 201, 202, 207, 209, 210, 211, 212, 213, 214, 217, 218

Breslau, 4, 16

Bromskirchen, 197

Brussels, 111, 122, 126

Buchenwald, 50, 71, 72, 88

Buhle, Walter, 54, 95, 98, 111

Bumper, 212, 213

Burgsteinfurt, 124, 125, 126, 158, 160, 161, 173, 174, 180, 181, 194
BzbV Heer, 74, 79, 193

Calais, 47, 50, 60, 65, 72, 74, 76, 80, 110, 114, 119
Cape Kennedy, 20
Celle, 96
Cement, 104
Cherbourg, 57, 72, 76, 77, 106, 107, 160
Cherwell, Lord, 61
Chiswick, 116, 117, 118
Churchill, Winston, 33, 57, 58, 59, 61, 62, 119, 127
CIU, 58
Cologne, 38, 126, 190
Crossbow, 77, 78, 92, 93, 94, 103, 106, 110, 111, 119, 127, 157, 214
Crüwell, Ludwig, 58
Cupola, 75, 101
Cuxhaven, 208, 209, 210, 212

D-Day, 2, 77, 78, 93, 99, 102, 214, 215
Danzig, 20, 45, 156
Dalfsen, 181, 182, 183, 184, 185
Darfeld, 124, 181
De Ceunynck, Simone, 164
De Keyserlei, 166
Degenkolb, Gerhard, 47, 48, 49, 52, 62, 63, 66, 71, 109, 139
DEMAG, 47, 90, 139
Denmark, 27
Dernau-Marienthal, 64
Dietrich, Sepp, 184
Division z.V., 113, 123, 185, 198
Doodlebug, 103
Dora, 71, 72, 86, 88, 140, 206
Dornberger, Walter, 1, 2, 7, 8, 9, 10, 11, 12, 13, 14, 15, 16, 17, 18, 19, 20, 22, 24, 25, 26, 27, 33, 34, 35, 36, 38, 40, 41, 42, 43, 44, 45, 46, 47, 48, 49, 51, 53, 54, 55, 56, 58, 63, 64, 65, 66, 67, 68, 69, 70, 74, 75, 79, 80, 81, 85, 86, 94, 95, 97, 98, 99, 104, 106, 109, 111, 112, 124, 140, 158, 173, 174, 201, 202, 203, 209, 210, 214, 217
Dorsch, Xavier, 75
Draayer, Wieger Jurjen, 132
Duindigt, 119, 136, 175, 186, 187
Dunkirk, 27, 72

East Anglia, 129, 133, 134
East Prussia, 34, 53, 75, 94
Ebensee, 104, 203
Eckener, Hugo von, 36
Eelerberg, 135, 146, 148, 150, 185
Eglin Field, 92, 93
Eisenhower, Dwight D., 78, 93, 102, 107, 122, 157, 198, 208, 215
Elbe River, 63, 197, 198, 199, 200, 201, 212
Elefant, 220, 221
English Channel, 27, 33, 47, 53, 60, 75, 77, 78, 80, 81, 102, 119, 174, 175
Entwicklungswerk (Development and Experimental Works), 21, 41
Eperlecques, 50, 51, 58, 65, 72, 73, 75, 76, 77, 80, 100, 106
Epping, 118
Engel, Rolf, 45
Euskirchen, 113, 126
Experimental Station West, 8, 9, 13

F1, 21, 50
Fallersleben, 110
Fallingbostel, 194, 195
Feuerleitpanzer (Firing Control Vehicle), 64, 109, 147, 148, 149, 150, 151, 183, 184, 196, 222, 224, 225, 226, 227
Fi-103, 46
Field Store, 84, 138, 142, 150, 176, 177, 183, 187, 220, 221, 227
Fieseler, 46, 81
Firing Table, 84, 91, 113, 114, 115, 119, 132, 137, 138, 142, 147, 150, 152, 156, 181, 188, 222, 223, 224, 226, 227
Fleisher, Karl, 207
Flying Bomb, 36, 39, 46, 53, 88, 93, 101, 102, 103, 105, 110, 114, 119, 120, 146, 154, 161, 169, 170, 172, 194
Freya, 64
Friedrichshafen, 36, 49, 50, 62, 63, 72
Fries Strabo Crane, 177, 180, 211, 220
Fritsch, Werner von, 13, 18, 19
Fromm, Friedrich, 34, 45, 46, 74, 99, 104
Frozen Lightning, 42, 153
FX Meiller Corporation, 64
FZG 76, 46

Garmisch-Partenkirchen, 207
Gehlert, 156, 181, 188
Gestapo, 89, 94, 95, 104, 203

Ghent, 102, 111, 170, 171
Goddard, Robert H., 5, 6, 13, 15, 34, 204, 205
Goebbels, Josef, 52, 53
Göring, Hermann, 62, 173, 202
Greifswalder Oie, 14, 17, 18, 24, 41, 97
Gröttrup, Helmut, 211, 212
Grossendorf, 45
Group North, 112, 113, 120, 121, 177, 191
Group South, 112, 113, 126, 127, 156, 196
Grünow, Heinrich, 10

Haagsche Bosch, 137, 142, 176, 186
Hachenburg, 126, 180, 188, 189, 196
Hall 41, 88, 139
Hamburg, 63, 64
Hanomag SS100 Tractor, 145, 180, 221, 222, 223
HARKO 191, 74, 81, 111
Harris, Arthur, 63
Hartman, Kenneth, 162, 166, 167
Haus Ingeborg, 201
Hayward, Greg, 161, 162, 163, 165
Heek, 125, 181, 194
Heeresversuchsanstalt Peenemünde (Army Research Center Peenemünde), 15
Heidekraut, 109, 110, 124, 125, 138, 148, 173, 201
Heidelager, 70, 83, 84, 85, 86, 90, 91, 96, 97, 98, 99, 108, 113, 114, 127, 128, 201
Heinemann, Erich, 80, 81, 84, 93, 103, 111, 113
Heinkel, 12, 22, 50, 53, 90
Helferskirchen, 126
Hellendoorn, 146, 147, 148, 150, 152, 153, 154, 161, 181, 185, 187, 190, 191, 192, 194, 195
Henschel, 49, 63, 71
Hermann, Rudolf, 17, 18, 41
Hermeskeil, 139, 156, 161
Hess, Rudolf, 19
Hessum, 182, 183
Heylandt, 11
Hillscheid, 156, 181
Himmler, Heinrich, 9, 19, 26, 44, 45, 46, 50, 51, 52, 69, 70, 71, 75, 81, 82, 94, 95, 96, 104, 108, 111, 112, 120, 202
Hitler, Adolf, 1, 2, 3, 7, 9, 10, 18, 19, 20, 22, 26, 27, 28, 33, 34, 35, 36, 38, 47, 48, 51, 52, 53, 54, 55, 56, 62, 63, 64, 58, 70,

72, 74, 75, 79, 80, 94, 95, 96, 99, 103, 104, 105, 106, 107, 111, 114, 121, 137, 140, 146, 147, 157, 168, 171, 172, 173, 187, 189, 191, 196, 200, 201, 202, 205, 215, 217, 218
Hoboken, 162, 169
Hochwald, 69, 70, 94
Hohmann, Walter, 4, 80, 113, 116, 175
Holste, Rudolf, 196, 197
Hook of Holland, 117, 142, 143, 176, 180
Horstig, Ritter von, 7, 8
Huntsville, 212
HVP, 27
Hydra, Operation, 67

Indigostraat, 174

Jodl, Alfred, 54, 95, 111
Jones, Reginald V., 25, 58, 59, 60, 61, 62, 118, 127
Junkers, 12, 88, 90

Kaltenbrunner, Ernst, 69, 95
Kammler, Hans, 70, 71, 86, 88, 105, 108, 109, 111, 112, 113, 118, 120, 121, 123, 124, 127, 129, 134, 140, 148, 173, 174, 184, 190, 192, 193, 198, 201, 202, 203, 207, 210
Kapo, 89
Karlshagen, 15, 27
Keitel, Wilhelm, 32, 54, 55, 95, 111
Kesselring, Albert, 14
Kesselwagen, 177, 184, 223
Kettler, Kurt, 71
Kippenburg, 128, 129
Kirburg, 188
Kleinbodungen, 90, 198
KNW, 50
Kochel, 74
Kohnstein, 62, 71, 86
Königsbrück, 12
Korolev, Sergei, 210
Korsett, 36
Köslin, 84, 85, 98, 124, 140, 148
Krakow, 83
Kummersdorf, 8, 9, 10, 11, 12, 13, 14, 15, 16, 19, 20, 21, 22, 45, 52, 58
Kunze, Heinz, 159
Kursk, 52
Kurzweg, Hermann, 18, 41

KZ Mittelbau, 71

Laura, 72
Leeb, Emil, 29, 44, 45
Leese, 194, 197
Legden, 124, 125, 126, 174
Lehesten, 72
Leiden, 143, 176, 177, 180, 194
Lenzen, 198, 199, 200
Lindemann, Frederick, 61
Lodz, 110
London, 2, 25, 33, 34, 38, 53, 57, 58, 59, 60,
   61, 71, 72, 77, 99, 101, 102, 103, 105,
   110, 111, 113, 114, 116, 117, 118, 119,
   120, 121, 122, 123, 128, 129, 134, 135,
   136, 139, 140, 143, 144, 145, 146, 148,
   154, 157, 170, 173, 174, 177, 187, 194,
   203, 208, 210, 214, 215, 216
Loosduinen, 174, 176, 180
Lübeck, 38
Ludendorff Bridge, 190, 191
Luftwaffe, 12, 14, 15, 22, 27, 33, 34, 38, 39,
   46, 47, 52, 56, 60, 62, 63, 64, 69, 172,
   189, 216, 217
Lulof, Cor, 152, 185
Luttenberg, 153, 154
LXV Army Corps, 80, 81, 84, 93, 98, 102,
   106, 111, 112, 113, 121, 140

Manteuffel, Hasso von, 172
Market Garden, Operation, 123, 124, 125,
   129, 134, 215
Marshall Space Flight Center, 213
Mataram, 183, 184
Mathieson, James, 167, 168
Mazuw, Emil, 45, 46
Me-163 Komet, 12
Medmenham, 60
Meillerwagen (V-2 Erector Trailer), 37, 64,
   84, 115, 122, 126, 129, 135, 142, 144,
   150, 153, 156, 188, 196, 211, 221, 222,
   223, 224, 226, 227
Merzig, 126, 127, 139, 161, 162
Metz, Richard, 81, 98, 99, 111, 120
Miesel, Johannes, 146, 184, 185
Milch, Erhard, 39
Mimoyecques, 76
Ministry of War, 47
Mirak, 4, 5
Mittelwerk, 71, 72, 74, 81, 86, 88, 89, 90,

91, 92, 97, 98, 108, 121, 129, 139, 140,
   144, 159, 180, 192, 198, 201, 205, 206,
   207, 209, 220
Montgomery, Bernard, 122, 123, 124, 135,
   170, 172, 194, 215
Moscow, 36, 128
Munich, 3, 64
Munitions Ministry, 28, 35

Nebel, Rudolf, 4, 5, 6, 7, 8, 9
Nebelwerfer, 12, 195, 196, 201
New Cross Shopping Center, 143
Niedersachswerfen, 62, 71, 90
Nijmegen, 113, 123
Nijverdal, 150
Norden Bombsight, 216
Nordhausen, 62, 86, 90, 107, 121, 140, 148,
   205, 206, 207, 208, 209, 210, 211
Normandy, 72, 76, 101, 102, 105, 109, 114,
   161, 214
Norway, 25, 27, 59
Norwich, 129, 133
Nuremburg, 35, 49, 107, 203

Oberammergau, 93, 201
Oberjoch, 201
Oberth, Hermann, 3, 4, 6, 7
Ockenburgh, 135, 142, 174, 180
OKW, 19, 26
Ommen, 150, 151, 183
Opel Blitz, 64, 182, 220, 223, 227
Opel, Fritz von, 4
Oslo Report, 25, 57, 59
Ostyn, Charles, 162, 163, 164, 166, 168, 169
Overcast, Operation, 208
Overlord, Operation, 77, 78, 92, 93, 102

P-1 (Prüfstand I), 21
P-7 (Prüfstand VII), 16, 20, 21, 36, 39, 41,
   43, 45, 57, 58, 63, 85, 90, 97, 109, 110
Paperclip, Operation, 208, 212
Paris, 81, 110, 111, 113, 115, 120, 127, 170,
   206, 207
Paris Gun, 7, 14
Pas de Calais, 47, 60, 72, 74, 76, 80, 101,
   110, 114, 119
Paulus, Friedrich von, 52
Peenemünde, 13, 14, 15, 16, 17, 18, 19, 20,
   22, 23, 24, 25, 26, 27, 28, 29, 33, 34, 35,
   36, 39, 40, 44, 45, 46, 47, 48, 49, 50, 51,

52, 53, 54, 55, 56, 57, 58, 59, 60, 61, 62,
   63, 64, 65, 66, 67, 68, 69, 70, 71, 72, 73,
   74, 79, 81, 82, 83, 84, 85, 86, 90, 91, 92,
   94, 95, 97, 104, 105, 109, 110, 111, 112,
   113, 119, 133, 140, 144, 148, 173, 174,
   188, 192, 193, 201, 203, 207, 208, 209,
   210, 211, 212, 213, 218
Penguin, Operation, 98, 111
Petites-Tailles, 113
Pilsen, 203
Pionierspark, 177
Plateau des Tailles, 113
Poland, 20, 27, 59, 70, 71, 79, 81, 83, 85,
   99, 100, 105, 124, 127, 138, 156, 196

Raaphorst, 123
Raderach, 72
RAF, 33, 38, 39, 57, 58, 59, 61, 62, 63, 67,
   68, 69, 72, 76, 78, 80, 93, 94, 99, 100,
   106, 114, 118, 123, 134, 135, 144, 161,
   167, 172, 175, 176, 177, 180, 183, 186,
   187, 191
RAF Fighter Command, 134
Raketenflugplatz, 3, 5, 6, 8, 10, 45
Rastenburg, 35, 53, 63, 64, 94, 104
Rax Werke, 49, 50, 71, 81
Red Army, 2, 36, 45, 108, 127, 192, 201
Redl-Zipf, 72
Redstone Arsenal, 212
Reichswehr (German Army), 7, 10, 13
Reinickendorf, 5, 6, 8, 10
Remagen, 187, 188, 189, 190, 192, 196
Repulsor, 5, 6
Rex Cinema, 166, 167, 168
Richthofen, Wolfram von, 12, 14
Rickhey, Georg, 158
Riedel, Klaus, 5, 16, 30, 64, 94, 95
Rijs, 128, 129, 132, 133, 134
Rijswijk, 136, 137, 142, 185
Robinson, Verne W., 165
Roquetoire, 76, 107
Rossmann, Erwin, 97
Roßbach, 126, 127
Rostock, 14
Rudolph, Arthur, 10, 50, 66
Rügen, 13, 24, 67
Rundstedt, Gerd von, 80, 106

SA, 10
Sanders, T. R. B., 127, 128, 209

Sandys, Duncan, 58, 59, 60, 61, 62, 76, 77,
   114, 118
Sant, Arthur van't, 174, 175
Sarnaki, 96, 99
Saturn Rocket 213, 218
Sauckel, Fritz, 49, 50
Saur, Karl, 62, 70
Sawatzki, Alwin, 63, 88
Scheveningen, 137, 140, 142, 175
Schloss Egelborg, 126, 174
Schloss Varlar, 158
Schmiedebach, 72
Schneidemuehl, 84
Schotterwerk Nord West (Gravel Quarry
   North West), 75
Serooskerke, 121
Shooters Hill, 144
Siemens, 22, 24, 90
Siewczynski, Fritz, 138
Siracourt, 76, 110
SIS, 58
Ski Sites, 77, 78, 92, 93, 94
Skoda, 203
Smithfield Market, 187
SOE, 99
Sottevast, 77, 106, 107
Special A-4 Committee, 47, 48, 50, 52
Speer, Albert, 29, 35, 36, 38, 39, 46, 47, 49,
   52, 53, 54, 55, 56, 62, 63, 64, 70, 71, 75,
   95, 158, 159, 173, 201, 202, 217
Spitfire, 39, 59, 123, 143, 144, 175, 176,
   187, 192
Sputnik 1, 3
SS, 2, 9, 10, 26, 45, 50, 51, 67, 68, 69, 70,
   71, 81, 82, 83, 86, 88, 89, 94, 95, 96, 99,
   104, 107, 110, 111, 112, 139, 147, 148,
   172, 173, 181, 184, 185, 193, 200, 201,
   203, 206
SS Werfer Battery 500, 110, 112, 124, 146,
   148, 156, 160, 161, 173, 125, 154, 181,
   184, 185, 190, 192, 194, 195, 196, 197,
   201
Stahlknecht, Detmar, 48, 49, 66
Stalin, Josef, 36, 52, 127, 196, 212
Stauffenberg, Claus von, 104
Staveley Road, 117, 118
Staver, Robert, 206, 207
Stegmaier, Gerhard, 41, 51, 65, 66, 84, 98
Steinhoff, Ernst, 41, 51, 53, 66
Sterpigny, 115

Stettin, 14, 45, 84, 95
Steyr, (See Stromversorgungswagen)
STO, 75
Stolzenau, 197
Stromversorgungswagen (Power Supply Vehicle), 64, 151, 224
Suchom Lake, 109, 110
Suttrop bei Warstein, 124, 185, 202

Taifun, 86
Tallboy, 73, 106, 110
Teilewerkstatt (Parts Production Plant Peenemünde), 21
Teniers Square, 164, 165
Test Stand I, 29 (see also, P-1)
Test Stand VII, 37, 40, 45, 52, 74, 79, 91, 105, 109, 110 (see also, P-7)
Test Stand X, 109
The Blitz, 34
The Hague, 113, 116, 119, 120, 123, 124, 125, 126, 134, 135, 136, 137, 138, 139, 140, 142, 143, 144, 148, 157, 172, 174, 175, 176, 180, 181, 185, 186, 187, 191, 192, 194, 218
Thiel, Walter, 16, 26, 27, 29, 30, 40, 43, 44, 48, 66, 67, 68, 69, 105, 159
Thoma, Wilhelm von, 58
Tippmann, Kurt, 185
Todt, Fritz, 28, 29, 34, 35
Todt Organization, 28, 47, 50, 51, 74, 75
Toftoy, Holgar, 206
Training and Experimental Battery 444, 65, 79, 84, 85, 91, 108, 112, 113, 114, 115, 120, 121, 122, 128, 129, 132, 133, 134, 136, 137, 138, 139, 142, 148, 161, 173, 174, 196, 197
Training and Replacement Detachment 271, 84, 142
Trassenheide, 50, 59, 67, 68, 69, 70
Tuchel, 85, 109, 187

United States, 1, 3, 5, 13, 38, 92, 158, 204, 205, 206, 208, 209, 211, 212, 217, 218
United States 8th Air Force, 72, 73, 78, 80, 92, 100, 109, 110, 216
United States 9th Air Force, 78
Usedom, 14, 15, 19, 20, 23, 27, 39, 56, 59, 66, 67

V-weapons, 2, 56, 72, 76, 77, 80, 92, 105,
111, 135, 157, 162, 163, 169, 170, 171, 176, 201, 214, 215, 218
Valier, Max, 4, 5, 7, 10
Verden, 196
Versailles Treaty, 7, 18
VfR, 4, 5, 6, 7, 8, 9, 10, 11
Vidalwagen, 87, 109, 128, 177, 181, 183, 220, 221, 223, 227
VKN, 34, 35, 38, 64, 65, 69, 79
Voss, Wilhelm, 203

Wa Prüf 11, 16
Wac Corporal, 212
Wahmke, Kurt, 16
Walcheren, 120, 121, 122, 124, 128, 157, 160
Walter, Hellmuth, 12
Wassenaar, 113, 116, 117, 118, 119, 123, 124, 136, 137, 138, 142, 144, 145, 180
Wasserfall, 21, 105
Watten, 50, 51, 66, 73, 100
War Cabinet, 57, 58, 59, 60, 61, 62, 77, 93, 119
Warsaw, 70, 109
Warsitz, Erich, 12
Weber, Wolfgang, 84, 112, 126
Wehrmacht (German Armed Forces), 8, 10, 19, 148, 152, 172, 185, 189, 214, 221, 226
Weimar Republic, 3, 6
Wellmann, Heinz, 107, 108, 148, 150, 151, 153, 154, 184
Welmbüttel, 196
Weser River, 173, 197
Westerwald, 126, 138, 156, 161, 173, 174, 180, 187, 188, 189, 197
Wetzling, Wolfgang, 199, 200, 201
White Sands, 206, 211, 212
Wiener Neustadt, 49, 50, 62, 81
Wifo, 62, 71
Wils, Bill, 180
Wind Tunnel, 17, 18, 21, 22, 69, 74, 173
Window, 67, 69
Winkler, Johannes, 4, 7, 8
Winter, Rita, 136, 137, 185, 186
Wizernes, 75, 76, 101, 106, 107
Wolfsschanze, 53, 69
Wolgast, 14, 173
Woolworth's, 145

XXX Corps, 123, 124

Zanssen, Leo, 7, 11, 20, 36, 41, 43
Zeppelin, 36, 49, 71
Zossen, 22
Zwolle, 182

Acknowledgements

During the years of research for this book, I have made numerous acquaintances with accomplished individuals from all over the world who offered their support, encouragement, and expertise. I would like to thank everyone who so generously contributed to this work and conveyed support, particularly my best friend Ed Straten; if it were not for his assistance, none of this would have transpired.

I have been fortunate enough to share thoughts and information with some of the world's most knowledgeable V-2 researchers. Thanks to Mike Imhoff for his always superlative insight and for sharing his incomparable collection of material. For answering all the tough questions, I must thank two of the world's foremost experts on the V-weapons (in my opinion)—Detlev Paul and Henk Koopman. I am also indebted to Cor Lulof for his technical perception and to Bert Koopman for sharing his intimate knowledge of V-2 battery vehicles and equipment. Thanks goes to Dr. Ken Hartman for providing his personal photographs taken in Antwerp during the V-bomb campaign of 1944–1945. Thanks to Gerhard Helm for his assistance with Peenemünde particulars and to Charles Ostyn for his wonderful maps and vivid personal account of surviving the war in Antwerp. I appreciate the help provided by all past and present members of the International V-2 Research Group; it has been a pleasure to work with each of you. I would also like thank my publisher, Bruce Franklin, for his direction and moral support.

Above all, I am deeply gratified by the encouragement and understanding provided by my loving wife Rhonda.